ASSESSMENT RESOURCES

Algebra 1

HOLT, RINEHART AND WINSTON

A Harcourt Classroom Education Company

Austin · New York · Orlando · Atlanta · San Francisco · Boston · Dallas · Toronto · London

To the Teacher

Algebra 1 Assessment Resources contains blackline masters that provide the teacher with both traditional and alternative assessments, giving teachers a variety of choices to best suit their needs.

- **Quick Warm-Up: Assessing Prior Knowledge** and **Lesson Quiz** (one page per lesson) contains a short set of prerequisite exercises for each lesson followed by a quiz consisting of free-response questions.

- **Mid-Chapter Assessment** (one page per chapter) contains multiple-choice questions and free-response questions that assess the instruction in the first half of each chapter.

- **Chapter Assessment** (two forms per chapter)

  Form A is a two-page multiple-choice test that includes assessment questions for every lesson in the chapter.

  Form B is a two-page free-response test that includes questions for every lesson in the chapter.

- **Alternative Assessment** (two forms per chapter)

  Form A is a one-page authentic-assessment activity or performance-assessment activity that involves concepts from the first half of the chapter.

  Form B is a one-page authentic-assessment activity or performance-assessment activity that involves concepts from the second half of the chapter.

Photo Credit
Front Cover: (background), Index Stock Photography Inc./Ron Russel; (bottom), Jean Miele MCMXCII/The Stock Market.

Printed in the United States of America

ISBN 0-03-054222-7

2   3   4   5   6   7   066   02   01   00

Table of Contents

Diagnostic Test

Write the letter that best answers the question or completes the statement.

_____ 1. A submarine dove to a depth of 328 feet below sea level. It dove another 289 feet before rising 300 feet. What integer represents the new position of the submarine in relation to sea level?

    a. 261   b. -261 c. -317 d. -917

_____ 2. Subtract: $-14 - 14$.

    a. -28 b. 0   c. 28   d. 14

_____ 3. Alissa withdrew $30 from her savings account each week for 6 consecutive weeks. How would you write the total amount as an integer?

    a. 180   b. -180 c. -30 d. -36

_____ 4. Divide: $-60 \div 15$.

    a. -900 b. 75   c. 4   d. -4

_____ 5. Add: $193.7 + 0.482$.

    a. 193,218   b. 194.182   c. $194.1\overline{82}$ d. 93.3634

_____ 6. Subtract: $0.082 - 0.0124$.

    a. -0.0696 b. 0.696   c. 0.0696   d. 0.704

_____ 7. Add: $\frac{3}{8} + \frac{5}{12}$.

    a. $\frac{2}{5}$ b. $\frac{2}{3}$ c. $\frac{8}{20}$ d. $\frac{19}{24}$

_____ 8. Subtract: $\frac{3}{4} - \frac{2}{5}$.

    a. $\frac{1}{20}$ b. $\frac{7}{20}$ c. 1   d. $\frac{1}{4}$

_____ 9. Caleb earns $11.18 per hour and works 37.5 hours per week. How much are his total earnings per week?

    a. $419   b. $419.25   c. $416.25   d. $413.66

_____ 10. A pipe that is 8.2 meters long must be cut into pieces that are 0.004 meters long. How many pieces can be cut?

    a. 2050   b. 8   c. 205   d. 20.5

_____ 11. Multiply: $\frac{9}{16} \times \frac{5}{6}$.

    a. $\frac{2}{5}$ b. $\frac{15}{16}$ c. $1\frac{19}{48}$ d. $\frac{15}{32}$

Diagnostic Test

_____ 12. Divide: $\frac{5}{6} \div \frac{7}{12}$.

    a. $\frac{35}{72}$ **b.** $\frac{5}{72}$ **c.** $1\frac{3}{7}$ **d.** $\frac{6}{35}$

_____ 13. Which of the following is true?

    a. $210{,}329 > 201{,}932$ **b.** $20{,}871 < 18{,}720$

    c. $687{,}876 = 687{,}786$ **d.** $7717 > 7771$

_____ 14. Which of the following sets of numbers is ordered from greatest to least?

    a. 490,328; 432,809; 398,430; 423,089   **b.** 490,328; 432,809; 423,089; 398,430

    c. 398,430; 423,089; 432,809; 490,328   **d.** 432,809; 490,328; 423,089; 398,430

_____ 15. Which of the following is true?

    a. $\frac{1}{3} < \frac{1}{4}$ **b.** $\frac{2}{3} < \frac{5}{12}$ **c.** $\frac{25}{30} = \frac{5}{6}$ **d.** $\frac{1}{4} < \frac{1}{5}$

_____ 16. How would you write $3\frac{4}{7}$ as an improper fraction?

    a. $3\frac{4}{7}$ **b.** $\frac{21}{7}$ **c.** $\frac{14}{7}$ **d.** $\frac{25}{7}$

_____ 17. Ellison threw a football 35 feet. Ethan threw the same football 10 yards. How much farther did Ellison throw the football?

    a. 19 feet   **b.** 5 feet   **c.** 30 feet   **d.** 11 feet

_____ 18. Solve the proportion $\frac{22}{x} = \frac{39}{156}$ for x.

    a. 88   **b.** 156   **c.** 1.7   **d.** 4

_____ 19. What are the next two terms of the sequence?

            17, 25, 33, 41, …

    a. 33 and 25   **b.** 49 and 57   **c.** 59 and 62   **d.** 47 and 56

_____ 20. Round 2.1985 to the nearest thousandth.

    a. 2.199   **b.** 2.198   **c.** 2.2   **d.** 2.19850

_____ 21. Evaluate $6(a - b)$ for $a = -3$ and $b = 2$.

    a. 11   **b.** 6   **c.** -30 **d.** 30

_____ 22. Express $\frac{1}{8}$ as a percent.

    a. 125%   **b.** 12.5%   **c.** 8%   **d.** 18%

Diagnostic Test

_____ 23. The volleyball team won 39 games, or 52%, of the games it played. How many games did it play?

    a. 80   b. 91   c. 75   d. 78

_____ 24. What is 6% of 150?

    a. 25   b. 12   c. 60   d. 9

_____ 25. 13 is what percent of 25?

    a. 12%   b. 50%   c. 52%   d. 1.92%

_____ 26. How would you graph the point $(-3, 1)$ on a coordinate plane?

    a. Start at the origin. Go 3 units to the right and 1 unit up.
    b. Start at the origin. Go 3 units to the left and 1 unit up.
    c. Start at the origin. Go 1 unit to the right and 3 units up.
    d. Start at the origin. Go 3 units to the left and 1 unit down.

_____ 27. Which inequality is graphed on the number line?

    a. $x \leq 2$ b. $x \geq 2$ c. $x = 2$ d. $x > 2$

_____ 28. What is the probability of rolling a number greater than 3 on one roll of a six-sided number cube?

    a. $\dfrac{1}{6}$ b. $\dfrac{1}{3}$ c. $\dfrac{1}{2}$ d. $\dfrac{3}{4}$

_____ 29. What is the mean of this set of data?

        89, 76, 78, 82, 89

    a. 89   b. 82   c. 78   d. 82.8

_____ 30. At a restaurant, there are 5 main entrees, 4 beverage choices, and 3 dessert choices. How many different dinners can be chosen?

    a. 12   b. 20   c. 60   d. 23

_____ 31. Rewrite the expression $3 \times 3 \times 3 \times 3$ in exponential form.

    a. 3^3 b. 3^4 c. 4^3 d. 3×4

Diagnostic Test

_____ **32.** What kind of transformation has been applied to the first figure to create the second figure?

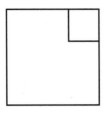 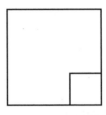

    a. translation   **b.** horizontal reflection   **c.** symmetrical   **d.** rotation

_____ **33.** The average length of bacteria is 0.0015 millimeters. How would you write this number in scientific notation?

    a. 1.5×10^3 **b.** 1.5×10^{-3} **c.** 1.5×10^{-1} **d.** 15×10^{-3}

_____ **34.** Which has the least value?

    a. $5 \times 6 \div 30$ **b.** $1 + 20 \div 20$ **c.** $18 \div 3 - 4$ **d.** $(6 + 4) \div 5$

_____ **35.** Write the phrase *7 less than some number* as a variable expression.

    a. $7 - x$ **b.** $7 \leq x$ **c.** $7 > x$ **d.** $x - 7$

_____ **36.** Solve: $b + 19 = 48$.

    a. $b = -29$ **b.** $b = 29$ **c.** $b = 19$ **d.** $b = 24$

_____ **37.** Solve: $16x = 320$.

    a. $x = 20$ **b.** $x = 32$ **c.** $x = 80$ **d.** $x = 8$

_____ **38.** Solve: $72 = 6n$.

    a. $n = 8$ **b.** $n = 7$ **c.** $n = 10$ **d.** $n = 12$

_____ **39.** Solve: $3y - 9 = 12$.

    a. $y = 4$ **b.** $y = 5$ **c.** $y = 7$ **d.** $y = 8$

_____ **40.** The circle graph shows how a senior class spent $4000. How much was spent on the school yearbook?

    a. $1600   **b.** $1000

    c. $40   **c.** $160

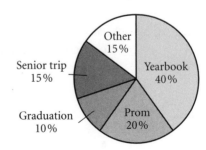

Quick Warm-Up: Assessing Prior Knowledge
1.1 *Using Differences to Identify Patterns*

Find each difference.

1. $11 - 5$ _____

2. $9 - 14$ _____

Evaluate each expression.

3. 7^2 _____

4. 5^3 _____

Lesson Quiz
1.1 *Using Differences to Identify Patterns*

Find the next two terms of each sequence.

1. $62, 68, 80, 98, \ldots$ _____

2. $125, 122, 116, 107, \ldots$ _____

3. $8, 16, 32, \ldots$ _____

4. $12, 19, 26, 33, \ldots$ _____

5. $6, 4, 10, 8, \ldots$ _____

6. $187, 178, 170, 163, \ldots$ _____

7. The first four terms of a sequence are 39, 41, 45, and 51. Find the first and second differences. What are the next three terms?

_____

8. Kayla trained with a pro to improve her bowling scores. During a tournament, she bowled a 162, 164, 168, and 174. If this pattern continues, what will be the scores of her next two games?

_____

Quick Warm-Up: Assessing Prior Knowledge
1.2 Variables, Expressions, and Equations

Evaluate each expression for *x* = 2 and *x* = 3.

1. $3x - 1$ _____   2. $4(x + 3)$ _____

Lesson Quiz
1.2 Variables, Expressions, and Equations

1. Given 1, 2, 3, 4, and 5 as values of b, find the values of $4b$. Show your work in a table.

Use guess-and-check to solve each equation.

2. $8w - 120 = 152$ _____   3. $16k - 7 = 137$ _____

4. $26c + 4 = -48$ _____   5. $87c + 2 = 31$ _____

6. If tickets for a concert cost $14 each, how many can you buy with $85?

_____

7. If 5 people want to split the cost of a $9 pizza equally, how much should each pay?

_____

8. Write an equation that models how many $2.98 videotapes can be bought for $15.

_____

Quick Warm-Up: Assessing Prior Knowledge
1.3 *The Algebraic Order of Operations*

Evaluate each expression.

1. $187 - 18.7$ _____

2. $30 + 3.2 + 0.3$ _____

3. 3^3 _____

4. $40 \cdot 12$ _____

Lesson Quiz
1.3 *The Algebraic Order of Operations*

Evaluate each expression.

1. $5 \cdot 2 + 7 \cdot 3$ _____

2. $1.5 \cdot 4.3 + 0.9$ _____

3. $6 \cdot 7 + 4$ _____

4. $\dfrac{2 + 3}{5 + 3 \cdot 5}$ _____

5. $\dfrac{5^2 - 2^4}{3}$ _____

6. $\dfrac{(6 + 8)(7 + 3)}{2 + 3 \cdot 4}$ _____

7. $19 - 5 \cdot 2$ _____

8. $4(6 - 5) + 3$ _____

9. $9 + 3(5 + 6)$ _____

10. 8^3 _____

11. $4(10)^3$ _____

12. $5^2 \cdot 3$ _____

Describe how you would evaluate each expression by using the order of operations. Then solve.

13. $(7 + 3) \div 2$ _____

_____

14. $5 \cdot (2 + 3) - 7$ _____

_____

15. $(6 \div 3) \cdot 2 + 7$ _____

_____

Mid-Chapter Assessment

Chapter 1 (Lessons 1.1–1.3)

Write the letter that best answers the question or completes the statement.

_____ 1. What are the next three numbers of the sequence?

25, 29, 34, 40, . . .

    a. 44, 48, 52   b. 36, 32, 28

    c. 47, 55, 64   d. 47, 53, 59

_____ 2. What is the first difference of this sequence?

70, 64, 58, 52, 46

    a. 6   b. -6 c. 40   d. 76

_____ 3. Which is an example of an equation?

    a. $2(l + w)$ b. $a^2 + 2 + c$ c. xy d. $x + y = 7$

_____ 4. If concert programs cost \$12, what is the cost of p programs?

    a. $12p$ b. $\dfrac{12}{p}$ c. $12 - p$ d. $12 + p$

5. Describe the pattern in 24, 72, 216, 648, . . . _____

6. Jamal needs \$650 to buy a camcorder. His savings passbook shows the following balances:

Date	Balance
January 1	\$100.00
February 1	150.00
March 1	250.00
April 1	400.00

If the pattern continues, in what month will he be able to buy the camcorder? _____

7. The fifth term of a sequence is 32. The first differences are 3. What are the first four terms? _____

8. Find the value of $6y$ by substituting for y. Make a table of values to show the substitutions of 1, 2, 3, 4, 5 for the variable. _____

Evaluate each expression.

9. $1.8 - 0.4 \cdot 2$ _____

10. $\dfrac{(3 + 2) - (4 \cdot 1)}{2 + 1 \cdot 2}$ _____

Quick Warm-Up: Assessing Prior Knowledge
1.4 Graphing With Coordinates

Graph each point on the same number line.

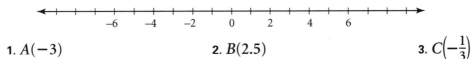

1. $A(-3)$ **2.** $B(2.5)$ **3.** $C\left(-\frac{1}{3}\right)$

Lesson Quiz
1.4 Graphing With Coordinates

Graph each list of points. State whether they lie on a straight line.

1. $A(1, 6), B(3, 7), C(5, 9)$

_____

2. $D(8, -2), E(7, -3), F(3, -7)$

_____

3. $P(-4, -1), Q(-6, 2), R(-4, 4)$

_____

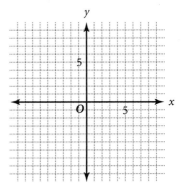

Determine the coordinates of the given points.

4. A _____

5. B _____

6. C _____

7. D _____

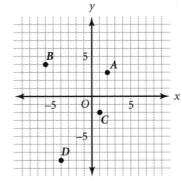

8. Given the equation $y = 3x - 2$ and 1, 2, 3, 4, and 5 as values of x, find the values of y. Make a table.

Quick Warm-Up: Assessing Prior Knowledge
1.5 *Representing Linear Patterns*

Find the next three terms of each sequence.

1. 11, 17, 23, 29, 35, …   2. 76, 68, 60, 52, 44, …   3. 1, 5, 13, 29, 61, …

Find the values for *y* by substituting 2, 4, 6, 8, and 10 for *x*.

4. $y = 2x - 5$ 5. $y = 3x + 1$

Lesson Quiz
1.5 *Representing Linear Patterns*

Write an equation to represent each data pattern.

1. _____

x	0	1	2	3	4	5	6
y	−3	2	7	12	17	22	27

2. _____

Number of books	0	1	2	3	4	5	6
Total cost ($)	5	7	9	11	13	15	17

Make a table of values for each equation, using 1, 2, 3, 4, and 5 as values for *x*. Draw a graph for each equation by plotting points from your data set.

3. $y = 7x$ 4. $y = 40 + 2x$

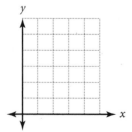   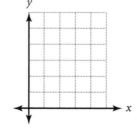

Quick Warm-Up: Assessing Prior Knowledge

1.6 Scatter Plots and Lines of Best Fit

Graph each point on the same coordinate plane.

1. $A(15, 700)$ **2.** $B(10, 350)$ **3.** $C(25, 300)$

 Lesson Quiz

1.6 Scatter Plots and Lines of Best Fit

Identify each scatter plot as showing a positive correlation, a negative correlation, or no correlation.

1. 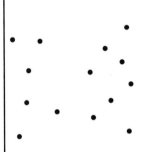   **2.** 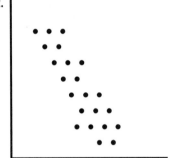   **3.**

_____   _____   _____

4. Given the scatter plot at the left below, which line better fits the points? Explain the reason for your choice.

   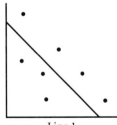   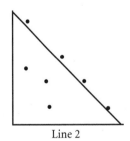

    Scatter plot   Line 1   Line 2

_____

_____

Chapter Assessment

Chapter 1, Form A, page 1

Write the letter that best answers the question or completes the statement.

_____ 1. What are the next three terms in this sequence?

$$12, 17, 22, 27, 32, \ldots$$

    a. 5, 10, 15   **b.** 36, 40, 44   **c.** 40, 45, 49   **d.** 37, 42, 47

_____ 2. What pattern is used to find the terms in 7, 16, 25, 34, 43, …?

    a. Multiply by 3.   **b.** Subtract 7.   **c.** Add 9.   **d.** Add 8.

_____ 3. Suppose that 7 homes are to be linked by underground cables, with 1 cable between each house. How many cables will there be?

    a. 7   **b.** 6   **c.** 8   **d.** 9

_____ 4. The first four terms of a sequence are 8, 14, 21, 29. What are the next three terms?

    a. 38, 49, 60   **b.** 39, 49, 59   **c.** 37, 54, 60   **d.** 38, 48, 59

_____ 5. The first three terms of a sequence are 3, 7, 14. The second differences are a constant 3. What are the next three terms of the sequence?

    a. 25, 29, 36   **b.** 17, 20, 23   **c.** 11, 7, 3   **d.** 24, 37, 53

_____ 6. What is the solution to the equation $3b + 5 = 41$?

    a. $b = 11$ **b.** $b = 12$ **c.** $b = 4$ **d.** $b = 5$

_____ 7. Which is an example of an expression?

    a. $8 > 1$ **b.** $3x + 1 = 7$ **c.** $19 + 7 = x$ **d.** $x + 5$

_____ 8. What is the value of $7n$ if n has a value of 4?

    a. $7n + 4$ **b.** 11   **c.** 28   **d.** 21

_____ 9. If blank videotapes cost \$6 each, what is the cost of t videotapes?

    a. $6t$ **b.** $6 + t$ **c.** $\frac{6}{t}$ **d.** $6 - t$

_____ 10. How many \$12 CDs can you buy with \$80?

    a. 6   **b.** 7   **c.** 8   **d.** 5

_____ 11. Evaluate $82 + 27 \cdot 35$.

    a. 2897   **b.** 144   **c.** 945   **d.** 1027

Chapter Assessment

Chapter 1, Form A, page 2

_____ 12. Evaluate $\frac{65 \cdot 32}{5 \cdot 16}$.

   a. 28   b. 2275   c. 26   d. 28.47

_____ 13. When evaluating the expression $16 - 4 \cdot 3 + 7^2$, which would you perform first?

   a. $16 - 4$ b. $3 + 7^2$ c. $3 + 7$ d. 7^2

_____ 14. Evaluate $b^2 - c^2$ for $b = 7$ and $c = 4$.

   a. 3   b. 33   c. 49   d. 16

_____ 15. What is the value of $(a + b) \cdot c$ for $a = 16$, $b = 7$, and $c = 5$?

   a. 115   b. 51   c. 23   d. 87

Use the graph to answer Exercises 16 and 17.

_____ 16. What are the coordinates of point A?

   a. $(5, 2)$ b. $(1, 4)$

   c. $(2, 5)$ d. $(5, 1)$

_____ 17. What are the coordinates of point C?

   a. $(3, 2)$ b. $(-3, 2)$

   c. $(-2, -3)$ d. $(-3, -2)$

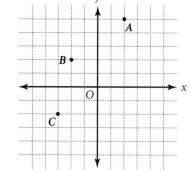

_____ 18. How would you describe the data in the scatter plot?

   a. strong correlation   b. positive correlation

   c. negative correlation d. little or no correlation

_____ 19. Which statement best describes the scatter plot?

   a. There is a positive correlation between age and annual earnings.
   b. As a person gets older, his or her earnings always increase.
   c. As a person gets older, his or her earnings decrease.
   d. You cannot tell whether there is a correlation between age and annual earnings.

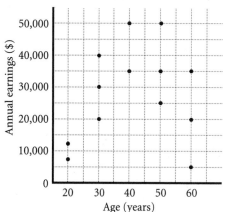

Chapter Assessment
Chapter 1, Form B, page 1

Find the next three numbers in each sequence.

1. 74, 65, 56, 47, . . . _____

2. 19, 22, 18, 21, . . . _____

3. 41, 48, 46, 53, . . . _____

4. The first four terms of a sequence are 37, 41, 47, and 55. Find the first and second differences. What are the next three terms?

_____

Make a table to show the substitutions of 1, 2, 3, 4, 5 for the variable.

5. Find the values of $7z$ by substituting for z.

6. Find the values of $3y$ by substituting for y.

Use guess-and-check to solve each equation.

7. $9w + 132 = 168$ _____   8. $12b - 13 = 59$ _____

9. $22x - 72 = 104$ _____   10. $51a + 75 = 279$ _____

Evaluate each expression.

11. $6 \cdot 5 + 3 \cdot 2$ _____   12. $9 \cdot 3 + 4 \div 3$ _____

13. $0.7 \cdot 2.4 + 7$ _____   14. $6.9 \cdot 4.2 + 6.7 \cdot 5$ _____

For $a = 7$, $b = 5$, and $c = 2$, evaluate each expression.

15. $a^2 + b^2$ _____   16. $a^2 + b - c$ _____   17. $(a + c) \cdot b$ _____

Chapter Assessment
Chapter 1, Form B, page 2

Graph the ordered pairs. State whether they lie on a straight line.

18. $(1, 5), (3, 7), (4, 8)$ _____

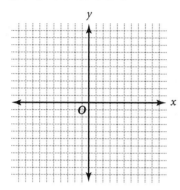

19. $(22, 25), (23, 24), (24, 23)$ _____

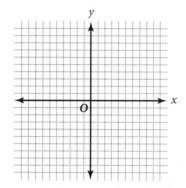

Find the values for _y_ by substituting 1, 2, 3, 4, 5 for _x_. Make a table.

20. $y = x + 5$

21. $y = 4x$

Describe each correlation as positive, negative, or little to none.

22.

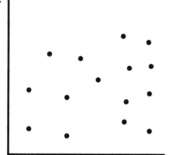

_____

23.

_____

24.

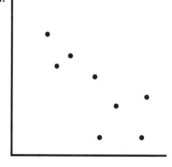

_____

25. Given the scatter plot shown at the left below, which line better fits the data? Explain the reason for your choice.

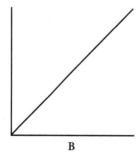

A

B

Alternative Assessment

The Snowflake Sequences, Chapter 1, Form A

TASK: Find the terms of numerical sequences by exploring the construction of a snowflake.

HOW YOU WILL BE SCORED: As you work through the task, your teacher will be looking for the following:

- whether you understand the concept of numerical sequences
- whether you are able to continue the pattern
- whether you can visualize and sketch the snowflake at different stages

The snowflake sequences were first produced by Swedish mathematician Helge Von Koch in 1904. The snowflake begins as an equilateral triangle, and a new equilateral triangle is added to the middle third of each side. Each successive figure is made exactly the same way using the new sides of the preceding figure. The figures show the beginning stages of a snowflake. Notice that the addition of each new equilateral triangle produces four new sides with a length of $\frac{1}{3}$ of a unit. As the steps continue, each side is divided into 4 parts. This can be continued indefinitely in the same manner.

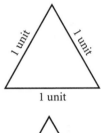

1. Make a rough sketch of the snowflake at stages 3–5. Then complete the table.

	Number of Sides	Length of Each Side	Perimeter of Snowflake
Stage 1	3	1	3
Stage 2	12	$\frac{1}{3}$	4
Stage 3	_____	_____	_____
Stage 4	_____	_____	_____
Stage 5	_____	_____	_____

The snowflake construction produces three sequences. Transfer the data from the table to show the sequences below. Then find the next two terms of each sequence.

2. Sequence 1:   3, 12, _____

3. Sequence 2:   1, $\frac{1}{3}$, _____

4. Sequence 3:   3, 4, _____

SELF-ASSESSMENT: Given the first three terms of any of the snowflake sequences, describe how you would determine the next term.

Alternative Assessment
Do You Always Make Your Bed?, Chapter 1, Form B

TASK: Conduct a survey to see if there is a correlation between age or gender and whether a person makes his or her bed regularly. Work with a partner and display your findings in a scatter plot.

HOW YOU WILL BE SCORED: As you work through the task below, your teacher will be looking for how well you can do the following:

- Compare two sets of data.
- Make a scatter plot.
- Make predictions and draw conclusions from data in a scatter plot.

1. Based on the data in your scatter plot, predict the age or gender of the person who rarely makes his or her bed.

_____

2. Draw a line of best fit on the scatter plot. How did you decide where to draw the line?

_____

_____

3. Explain how you can determine whether there is a positive or negative correlation for the data in a scatter plot. What correlation is there for your scatter plot?

_____

_____

_____

4. Hypothesize by using your scatter plot. Do you think your data is typical of a larger population? Do you think it represents a trend? Explain.

_____

_____

_____

_____

SELF-ASSESSMENT: On a separate sheet of paper, explain what is most important to understand about scatter plots. Then describe the part(s) of the task above that gave you the most difficulty and how you worked through it.

Quick Warm-Up: Assessing Prior Knowledge
2.1 The Real Numbers and Absolute Value

Evaluate each expression.

1. $190 - 128 \div 4$ _____

2. $21 - 6 \cdot 3 + 4^2$ _____

3. $8 \cdot 9 + 7 \cdot 5$ _____

4. $2(5 + 8 - 6) + 11$ _____

5. $3^2 \cdot 4 + [9 - (2^2 - 1)]$ _____

Lesson Quiz
2.1 The Real Numbers and Absolute Value

Insert <, >, or = to make each statement true.

1. 7 _____ 17

2. 0 _____ -1

3. -0.3987 _____ 0.3984

4. $-1\frac{2}{3}$ _____ -1

5. -0.5 _____ -5

6. 8.26 _____ 8.32

Find the opposite of each number.

7. 3 _____

8. -7 _____

9. $-(-8)$ _____

10. 0 _____

11. -6.1 _____

12. $-(0.1 + 6.7)$ _____

Find the absolute value of each number.

13. -5 _____

14. 9 _____

15. 0 _____

16. 1.5 _____

17. $-(17)$ _____

18. $-\left(\frac{5}{6}\right)$ _____

Simplify each expression.

19. $-(16 - 3)$ _____

20. $|-32|$ _____

21. $-(-1)$ _____

Quick Warm-Up: Assessing Prior Knowledge
2.2 Adding Real Numbers

Find each sum.

1. $868 + 75$ _____

2. $\frac{5}{7} + \frac{1}{2}$ _____

3. $7.2 + 6.9$ _____

Find each difference.

4. $401 - 83$ _____

5. $\frac{5}{6} - \frac{3}{10}$ _____

6. $16 - 9.24$ _____

Lesson Quiz
2.2 Adding Real Numbers

Find each sum.

1. $7 + (-6)$ _____

2. $-14 + 9$ _____

3. $-5.3 + (-6.3)$ _____

4. $9 + (-3)$ _____

5. $4.6 + (-7.9)$ _____

6. $5 + (-6.5)$ _____

7. $10 + (-11)$ _____

8. $-\frac{1}{4} + \left(-\frac{1}{2}\right)$ _____

9. $1\frac{2}{3} + \left(-\frac{3}{4}\right)$ _____

10. $3 + (-1)$ _____

11. $-6 + (-13)$ _____

12. $7 + (-3.4)$ _____

13. $-2.5 + 3.2$ _____

14. $\left(\frac{3}{4}\right) + \left(-\frac{2}{3}\right)$ _____

15. $62 + (-62)$ _____

Let $a = -3$, $b = 3$, and $c = 4$. Evaluate each expression.

16. $a + c$ _____

17. $c + (-a)$ _____

18. $-b + (-a)$ _____

19. $c + (-a) + b$ _____

20. $-c + (-b) + (-a)$ _____

21. $(-b) + c + a$ _____

Quick Warm-Up: Assessing Prior Knowledge
2.3 *Subtracting Real Numbers*

Find the opposite of each number.

1. -9 _____

2. 3.7 _____

3. 0 _____

4. $-\dfrac{1}{3}$ _____

Find each sum.

5. $-5 + 8$ _____

6. $-13 + (-6)$ _____

7. $-\dfrac{3}{5} + \dfrac{2}{5}$ _____

8. $6.9 + (-4.28)$ _____

Lesson Quiz
2.3 *Subtracting Real Numbers*

Find each difference.

1. $3 - 6$ _____

2. $3 - 9$ _____

3. $-2 - (-3)$ _____

4. $-5.4 - (-2.6)$ _____

Find each sum or difference.

5. $-9 - (-9)$ _____

6. $-4 - 8$ _____

7. $-3 - (-5)$ _____

8. $-4.3 + 7.2 - 1.1$ _____

9. $-5 - (-2.8) - 0.21$ _____

10. $-6.2 + 8.1 - 1.9$ _____

Let $x = 3$, $y = -2$, and $z = -5$. Evaluate each expression.

11. $x + y - z$ _____

12. $y - (x + z)$ _____

13. $z - y - x$ _____

NAME _____ CLASS _____ DATE _____

Quick Warm-Up: Assessing Prior Knowledge
2.4 Multiplying and Dividing Real Numbers

Find each product.

1. $(3.8)(4.1)$ _____

2. $\frac{3}{5} \cdot \frac{1}{6}$ _____

Find each quotient.

3. $6.8 \div 0.4$ _____

4. $\frac{1}{10} \div \frac{1}{4}$ _____

Lesson Quiz
2.4 Multiplying and Dividing Real Numbers

Write the reciprocal of each number.

1. $\frac{2}{3}$ _____

2. $\frac{5}{4}$ _____

3. $\frac{3}{7}$ _____

4. $\frac{-2}{4}$ _____

5. $\frac{12}{-3}$ _____

6. 16 _____

7. -13 _____

8. $\frac{-22}{-2}$ _____

9. $\frac{4}{36}$ _____

10. $\frac{7}{-11}$ _____

11. $\frac{-18}{3}$ _____

12. $\frac{1}{-4}$ _____

Evaluate.

13. $(-1)(-5)$ _____

14. $(-4)(5)$ _____

15. $24 \div (-6)$ _____

16. $-2 \cdot 8$ _____

17. $-63 \div 9$ _____

18. $-144 \div (-12)$ _____

19. $(-4)(-4)(-4)$ _____

20. $-9 \div 3 \cdot (-2)$ _____

21. $-\frac{2}{3} \cdot \frac{3}{2}$ _____

22. $-\frac{8(-5)}{-4}$ _____

23. $-\frac{3}{5} \div \left(-\frac{3}{4}\right)$ _____

24. $\frac{-9 + -6}{-3}$ _____

Mid-Chapter Assessment

Chapter 2 (Lessons 2.1–2.4)

Write the letter that best answers the question or completes the statement.

_____ 1. Add: $3.72 + (-4.9)$.

        a. -8.62 **b.** 8.62 **c.** 1.18 **d.** -1.18

_____ 2. Evaluate $-5.7 + 4.5 - 2.2$.

        a. -3.4 **b.** 4.5 **c.** 3.4 **d.** 5.7

_____ 3. Multiply: $(-3)(-2)(-4)$.

        a. 24 **b.** -9 **c.** -24 **d.** -9

_____ 4. Evaluate $\dfrac{-10 + (-15)}{-5}$.

        a. 1 **b.** -5 **c.** 5 **d.** -1

_____ 5. Evaluate $xy + z$ for $x = 2$, $y = -3$, and $z = 5$.

        a. 1 **b.** -1 **c.** 11 **d.** -11

_____ 6. Evaluate $\dfrac{x - y}{y}$ for $x = \frac{1}{2}$ and $y = -\frac{3}{4}$.

        a. 3 **b.** $-\frac{1}{3}$ **c.** $-\frac{5}{3}$ **d.** $\frac{5}{3}$

_____ 7. Evaluate $\dfrac{3(-1) + (-7)}{20}$

        a. 10 **b.** -2 **c.** $\frac{1}{2}$ **d.** $-\frac{1}{2}$

Write the reciprocal of each number.

8. $\dfrac{6}{5}$ _____   9. $\dfrac{-2}{20}$ _____   10. $\dfrac{1}{3}$ _____

11. $\dfrac{2}{7}$ _____   12. $\dfrac{6}{3}$ _____   13. $\dfrac{11}{22}$ _____

Quick Warm-Up: Assessing Prior Knowledge
2.5 Properties and Mental Computation

Evaluate.

1. $44 + (36 + 32)$ _____   2. $3(4 + 2)$ _____   3. $5 - (6 + 2)$ _____

4. $2 \cdot (25 \cdot 3)$ _____   5. $4(9.5)$ _____

Lesson Quiz
2.5 Properties and Mental Computation

1. Complete each step and name the property used.

$25 \cdot (12 \cdot 4)$

$= 25 \cdot ($ _____ $\cdot 12)$ _____

$= ($ _____ $\cdot 4) \cdot 12$ _____

$=$ _____ $\cdot$ _____

$=$ _____

Use the Associative, Commutative, and Distributive Properties to find each sum or product.

2. $(42 + 18) + 29$ _____   3. $(25 \cdot 16) \cdot 4$ _____

4. $(29 \cdot 2) \cdot 5$ _____   5. $(578 + 416) + 122$ _____

Name the property illustrated. Be specific.

6. $78 + 89 = 89 + 78$ _____

7. $(18 + 9) + 25 = 18 + (9 + 25)$ _____

8. $20(6 + 9) = 20(6) + 20(9)$ _____

9. $58(16 - 5) = 58(16) - 58(5)$ _____

Quick Warm-Up: Assessing Prior Knowledge
2.6 Adding and Subtracting Expressions

Find each sum.

1. $12 + (-8)$ _____
2. $-6 + (-5)$ _____
3. $-11 + 7$ _____

Find each difference.

4. $4 - 9$ _____
5. $-7 - 8$ _____
6. $55 - (-20)$ _____

Lesson Quiz
2.6 Adding and Subtracting Expressions

Use the Distributive Property to show that the following are true statements:

1. $6w + 3w = 9w$ _____
2. $2t - t = t$ _____
3. $-5r - 2r = -7r$ _____

4. $3p + 6p = 9p$ _____
5. $\frac{1}{2}k + k = 1\frac{1}{2}k$ _____
6. $0.5s - 0.25s = 0.25s$ _____

7. $5q + (-2q) = 3q$ _____
8. $27g - 12g = 15g$ _____
9. $\frac{2}{5}k - \frac{4}{5}k = \frac{-2}{5}k$ _____

Simplify the following expressions.

10. $(2x + 5) + (4x + 3)$ _____

11. $(3b + 1) + (2b - 5)$ _____

12. $(6z - 3) + (4z - 5)$ _____

13. $(3.3x + 2.2y) + (3x - 0.6y)$ _____

14. $6x - (4 - 3x)$ _____

15. $(11d + 4e) - (6d + 3e)$ _____

16. $(7m + 6n) - (5m - 8n)$ _____

17. $(12p - 3q + 5r) - (8p + 2q - r)$ _____

Quick Warm-Up: Assessing Prior Knowledge
2.7 Multiplying and Dividing Expressions

Find each product.

1. $(-5)(18)$ _____

2. $(-9)(-7)$ _____

Find each quotient.

3. $16 \div (-8)$ _____

4. $\dfrac{-54}{-18}$ _____

Lesson Quiz
2.7 Multiplying and Dividing Expressions

Simplify the following expressions. Use the Distributive Property if needed.

1. $3 \cdot 7x$ _____

2. $-7x \cdot 3$ _____

3. $-7x \cdot 3x$ _____

4. $-77x \div 11$ _____

5. $6x \cdot 5 - 2x \cdot 3$ _____

6. $-36x \div 6x$ _____

7. $\dfrac{22 - 44x}{22}$ _____

8. $\dfrac{-15x + 10}{5}$ _____

9. $\dfrac{6 + 12x}{6}$ _____

10. Which of the expressions below are equivalent?

   a. $(-3x + 8) \div 4$ b. $(8 - 3x) \cdot \dfrac{1}{4}$ c. $(-8 + 3x) + 1 - 4$ d. $\left(\dfrac{1}{4}\right) - (-3x + 8)$

11. Which of the expressions below are equivalent?

   a. $\dfrac{3(2x - 1)}{3}$ b. $(2x - 1)$ c. $\dfrac{4(2x - 1)}{8}$ d. $\dfrac{2(3x - 1)}{2}$

12. Rachel wants to wallpaper one wall in her family room. The length of the wall is 14 feet and the height is 12 feet. How many square feet of wallpaper does she need?

   _____

Chapter Assessment

Chapter 2, Form A, page 1

Write the letter that best answers the question or completes the statement.

_____ 1. Find the opposite of $-(3 + 3.5)$.

    a. -6.5 b. 6.5 c. 0.5 d. -0.5

_____ 2. Add: $-25 + 26 + |-18|$.

    a. 32 b. -16 c. 19 d. 68

_____ 3. Juan has \$63 in his savings account. He withdrew \$16 one week and deposited \$28 the next week. What is the balance in his account?

    a. \$75   b. \$107   c. \$19   d. \$51

_____ 4. Complete: $-2 - (-7) = -2 + ?$.

    a. -7 b. -9 c. 7 d. 9

_____ 5. Subtract: $-3 - (-47)$.

    a. -50 b. 44 c. 50 d. -44

_____ 6. Evaluate $(-64)(-110)$.

    a. -7040 b. 46 c. 174 d. 7040

_____ 7. On Sunday, the temperature was 82°F in Phoenix and -7°F in Minneapolis. How much warmer was it in Phoenix that day?

    a. 75°F   b. -7°F   c. 82°F   d. 89°F

_____ 8. Evaluate $7b + 9$ for $b = \frac{1}{2}$.

    a. 23 b. $3\frac{1}{2}$ c. $12\frac{1}{2}$ d. $16\frac{1}{2}$

_____ 9. Which expression is equivalent to $(4)(6 - 2x)$?

    a. $24 - 2x$ b. $24 - 8x$ c. $3(8 - 2x)$ d. $6(4 - 2x)$

_____ 10. Sam charges \$0.89 per pound for grapes. Which expression shows the cost for p pounds of grapes?

    a. $0.89 \div p$ b. $0.89 + p$ c. $0.89p$ d. $0.89(9p - 1)$

———————— **11.** Add: $(2b - 4) + (8b - 6)$.

    a. $10b + 2$ **b.** $6b - 2$ **c.** $10b - 10$ **d.** $-6b + 10$

———————— **12.** Add: $(5e + 9f) + (2e - 4f)$.

    a. $7e + 13$ **b.** $7e - 5f$ **c.** $3e + 5f$ **d.** $7e + 5f$

———————— **13.** Let $a = -2$ and $b = 4$. Find $a + b$.

    a. -6 **b.** -2 **c.** 4 **d.** 2

———————— **14.** Subtract: $(6b + 1) - (2b + 4)$.

    a. $3b - 3$ **b.** $4b + 5$ **c.** $8b + 5$ **d.** $4b - 3$

———————— **15.** A recipe for clam chowder says to add water to the clam broth to get $2\frac{1}{3}$ cups of liquid. If there is $\frac{1}{2}$ cup of clam broth, how much water should be added?

    a. $\frac{1}{3}$ cup   **b.** $1\frac{5}{6}$ cup   **c.** $\frac{5}{6}$ cup   **d.** $\frac{1}{2}$ cup

———————— **16.** Identify the property illustrated by $4(3 + 2) = (4 \cdot 3) + (4 \cdot 2)$.

    a. Associative   **b.** Commutative   **c.** Distributive   **d.** Transitive

———————— **17.** Identify the property illustrated by $\frac{3}{4} + \frac{2}{3} = \frac{2}{3} + \frac{3}{4}$.

    a. Associative   **b.** Commutative   **c.** Distributive   **d.** Transitive

———————— **18.** Identify the property illustrated by $4 + (3 + 7) = (4 + 3) + 7$.

    a. Associative   **b.** Commutative   **c.** Distributive   **d.** Transitive

Chapter Assessment

Chapter 2, Form B, page 1

Find each sum.

1. $(27) + (-35)$ _____

2. $-64 + 61$ _____

3. $-3 + (-4)$ _____

Find each difference.

4. $(38) - (-19)$ _____

5. $62 - 83$ _____

6. $-13 - (-89)$ _____

Use mental math to find each sum or product.

7. $-30 + (-14) + 10$ _____

8. $(32 \cdot 5) \cdot 20$ _____

Evaluate.

9. $(-32)(-16)$ _____

10. $(-48) \div (-8)$ _____

Evaluate $4p + 7$ for the following values of p:

11. 30 _____

12. 5.5 _____

13. $\frac{1}{3}$ _____

Simplify the following expressions. Use the Distributive Property if needed.

14. $3 \cdot 4y$ _____

15. $-4y \cdot 3$ _____

16. $-12y \div (-3)$ _____

17. $\frac{10 - 22x}{10}$ _____

18. $\frac{-60x + 3.3}{6}$ _____

19. $24x \cdot 2x$ _____

20. In California, Mt. Whitney rises to 14,495 feet above sea level. Nearby, Death Valley sinks to 282 feet below sea level. What is the

difference in elevation between these two points? _____

21. What property is illustrated by the equation $m + n = n + m$? _____

22. Evaluate $|-35|$. _____

Simplify each expression.

23. $(3c + 7) + (4c - 3)$ _____

24. $(9x + 2) - (3x + 7)$ _____

Chapter Assessment

Chapter 2, Form B, page 2

Evaluate each expression for $a = 7$, $b = 5$, and $c = 2$.

25. $a + b$ _____

26. $a + b - c$ _____

27. $(a + c) \cdot b$ _____

Find the opposite of each expression. Then simplify if possible.

28. $8y + 3y$ _____

29. $3m - 4m$ _____

30. $-b - c$ _____

Perform the indicated operations.

31. $10x - 2x$ _____

32. $5d - (2 - d)$ _____

33. $(5r + 2s) + (6r + 4s)$ _____

34. $(8v - 7w) - (7v - 2w)$ _____

35. Make up a real-world situation that can be modeled by the equation $x + 19 = 45$.

_____

_____

_____

36. Ray bought 6 cartons of small bedding plants and 2 pots of full-grown plants. The store ran a sale the following week, so he purchased 3 additional cartons of the small plants and 4 pots of full-grown plants. Represent the cartons and pots algebraically, and

    determine the expression for the sum. _____

_____

_____

Jeffrey opened a savings account with a $30 deposit. He made 2 more deposits of $26 and 2 withdrawals of $15 and $18.

37. What is the total amount currently in Jeffrey's account? _____

38. What is the cumulative change in Jeffrey's account since he opened it? _____

Simplify each expression.

39. $-4(5x + 2)$ _____

40. $(4x)(3) + (5x)(2)$ _____

41. $x^2 - (5 - 2x^2)$ _____

Alternative Assessment

What's My Rule? Chapter 2, Form A

TASK: Add and subtract real numbers by using number lines and Addition and Subtraction Properties.

HOW YOU WILL BE SCORED: As you work through the task, your teacher will be looking for the following:

- whether you can describe the procedure for adding real numbers by using number lines
- whether you can make a generalization about adding real numbers without a number line
- whether you can explain and give examples of how you can subtract positive and negative real numbers by using addition

Adding real numbers can be modeled by using number lines.

1. Describe how you would use the number line to add −4 and 5.

_____

2. Show the addition of −4 and 5 on a number line in two ways.

3. Explain how you would add two numbers with the same sign without using a number line.

_____

_____

4. Explain how you would add two numbers with opposite signs without using a number line.

_____

_____

5. Show how you can subtract positive and negative numbers by using addition.

_____

6. Give an example which illustrates that subtraction of numbers is not commutative.

_____

SELF-ASSESSMENT: Explain what is meant by inverse operations and how this helps you to add and subtract numbers.

Algebra 1

Alternative Assessment
Hot, Hot, Hot! Chapter 2, Form B

TASK: Explore and interpret temperature data.

HOW YOU WILL BE SCORED: As you work through the task, your teacher will be looking for the following:

- whether you can add and subtract real numbers
- how well you can interpret data presented in a list and use it to solve problems

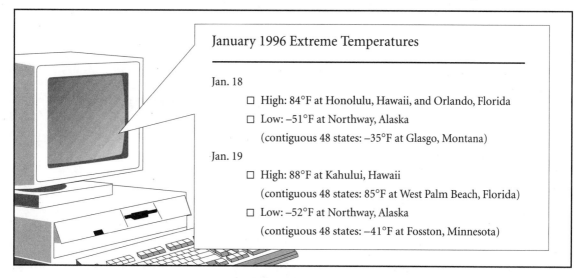

January 1996 Extreme Temperatures

Jan. 18
- ☐ High: 84°F at Honolulu, Hawaii, and Orlando, Florida
- ☐ Low: –51°F at Northway, Alaska
  (contiguous 48 states: –35°F at Glasgo, Montana)

Jan. 19
- ☐ High: 88°F at Kahului, Hawaii
  (contiguous 48 states: 85°F at West Palm Beach, Florida)
- ☐ Low: –52°F at Northway, Alaska
  (contiguous 48 states: –41°F at Fosston, Minnesota)

1. How many degrees warmer was Honolulu than Northway on January 18? _____

2. Justin says that the difference between the high and low temperatures was 5 degrees greater on January 19 than on January 18. Laura says the difference between the temperature ranges is 3 degrees. Who is right? How do you know?

_____

_____

3. What was the range of temperatures in the 48 contiguous states on January 19?

_____

4. How many degrees colder was Northway than Fosston on January 19? _____

5. On which day was the low temperature in the 48 contiguous states warmer? How much warmer?

_____

SELF-ASSESSMENT: Draw and label thermometers showing the temperature ranges for January 18 and for January 19.

Quick Warm-Up: Assessing Prior Knowledge

3.1 Solving Equations by Adding and Subtracting

Find each sum or difference.

1. $-10 + 4$ _____

2. $12 + (-12)$ _____

3. $14 - (-14)$ _____

4. $-8 - (-8)$ _____

Lesson Quiz

3.1 Solving Equations by Adding and Subtracting

Solve each equation. Check the solution.

1. $y + 7 = 15$ _____

2. $y - 7 = 8$ _____

3. $y + 15 = 7$ _____

4. $y - 15 = 7$ _____

5. $x - 3 = 10$ _____

6. $x - 10 = 3$ _____

7. $x + 3 = 10$ _____

8. $x + 10 = 3$ _____

9. $18 = b - 4$ _____

10. $-6 = c + 5$ _____

11. $y - \frac{1}{3} = \frac{1}{2}$ _____

12. $y - \frac{3}{5} = \frac{7}{10}$ _____

Write an equation for each situation. Then solve the equation.

13. The odometer of your car registered 49,116 miles before
    your 264-mile trip. What did the odometer register after
    the trip?

    _____

14. A pair of shoes cost $59.95 plus tax. If the change is $0.55
    from $65, how much was the tax?

    _____

Quick Warm-Up: Assessing Prior Knowledge
3.2 *Solving Equations by Multiplying and Dividing*

Find each product or quotient.

1. $(6)(-7)$ _____

2. $(-8)(-12)$ _____

3. $-148 \div (-4)$ _____

4. $-8 \div 16$ _____

5. $(-1.5)(0.3)$ _____

Lesson Quiz
3.2 *Solving Equations by Multiplying and Dividing*

Solve each equation and check your solution.

1. $8x = 53$ _____

2. $\frac{y}{42} = 7$ _____

3. $\frac{c}{4} = -1$ _____

4. $-3x = 45$ _____

5. $-14 = 8b$ _____

6. $6.6y = 44$ _____

7. $\frac{x}{-3} = 2.5$ _____

8. $x + \frac{2}{3} = 5$ _____

9. $555x = 1665$ _____

10. $x - 555 = 1665$ _____

11. $y + \frac{2}{3} = 6$ _____

12. $y + 0 = -473$ _____

13. $x - \frac{5}{6} = 4$ _____

14. $\frac{x}{30} = 2.3$ _____

15. $4a = 14$ _____

16. $x - 9 = 45$ _____

17. Rico paid $2.24 for four markers. How much did each one cost? _____

18. Donna wants to earn $120. If she earns $5 an hour babysitting, how many hours must she work to reach her goal?

_____

Quick Warm-Up: Assessing Prior Knowledge
3.3 *Solving Two-Step Equations*

Solve for *x*.

1. $3 + x = 11$

2. $x - 7 = 19$

3. $21 - x = 36$

4. $3x = 18$

_____ _____ _____ _____

5. $9x = 45$

6. $6x = 15$

7. $\frac{x}{3} = 4$

8. $\frac{x}{15} = 3$

_____ _____ _____ _____

Lesson Quiz
3.3 *Solving Two-Step Equations*

Solve each equation.

1. $5p - 3 = 22$ _____

2. $4x - 3 = 29$ _____

3. $6b + 12 = -18$ _____

4. $-7x - 15 = 6$ _____

5. $2x - 9 = 11$ _____

6. $12y - 15 = 9$ _____

7. $6 - 7r = 20$ _____

8. $15 - 8m = 23$ _____

9. $3.1 + 2m = 6.3$ _____

10. $7n - 8.5 = 4.1$ _____

11. $\frac{y}{8} - 6 = 2$ _____

12. $\frac{y}{9} - 6 = 3$ _____

13. $\frac{z}{10} + 6 = 2$ _____

14. $\frac{n}{3} + 3 = 5$ _____

15. A student council is planning to make posters for a major fundraiser. They will pay $5 to each person who helps. There is a $10 cost for materials. If they have $30 to spend on the posters, how many people can be hired to help? _____

16. The cost of concert tickets is $18.50 per ticket. There is also a one-time $5 service charge. If a person spent $116 buying tickets, how many concert tickets were purchased? _____

Mid-Chapter Assessment
Chapter 3 (Lessons 3.1–3.3)

Write the letter that best answers the question or completes the statement.

_____ 1. Solve $6x + 8 = -46$.

   a. $x = 9$ b. $x = -9$ c. $x = 6.3$ d. $x = 5$

_____ 2. Solve $4(y + 30) = 300$.

   a. $y = 45$ b. $y = -45$ c. $y = 37.5$ d. $y = -37.5$

_____ 3. Solve $8t + 3 - 2t + 11 = 58$.

   a. $t = 12$ b. $t = -12$ c. $t = \frac{22}{3}$ d. $t = -\frac{22}{3}$

_____ 4. Solve $2m - 4 = 8m$.

   a. $m = \frac{2}{3}$ b. $m = -\frac{2}{3}$ c. $m = -\frac{2}{5}$ d. not given

_____ 5. Solve $3n + 16 = 21n - 38$.

   a. $n = \frac{9}{4}$ b. $n = -3$ c. $n = \frac{11}{9}$ d. not given

_____ 6. Solve $6(x - 20) + 12x = -10x + 8$.

   a. $x = \frac{31}{7}$ b. $x = 16$ c. $x = -16$ d. not given

7. Sean works at a car dealership. He earns $300 a week plus 1% of his sales. What should his sales be in order to make $800 a week?

_____

8. Anita is going to mix 6 gallons of a solution that is 20% oil with a solution that is 44% oil. How much of the second solution should she add to the 20% oil solution to obtain a solution that is 32% oil?

_____

For a school activity, the seniors at Mountainview High School are having a Valentine's Day dinner. The total cost for the food is $55, and they must pay the school $0.50 per person for use of the cafeteria. They plan to charge $8 per person to attend.

9. Write an equation for the profit in terms of the number of people attending.

_____

_____

10. How many people must attend to make a profit of $500? _____

Quick Warm-Up: Assessing Prior Knowledge

3.4 Solving Multistep Equations

Solve for *x*.

1. $3 + 2x = 11$

2. $13x - 7 = 19$

3. $21 - 5x = 36$

_____

_____

_____

4. $3x + 6 = 18$

5. $9x + 8 = 45$

6. $6x + 17 = 28$

_____

_____

_____

Lesson Quiz

3.4 Solving Multistep Equations

Solve each equation.

1. $3p + 7 = 25$ _____

2. $8x - 9 = 31$ _____

3. $-5c + 17 = -28$ _____

4. $-7m + 23 = -5$ _____

5. $\frac{x}{7} + 5 = 10$ _____

6. $\frac{z}{10} = z - 40$ _____

7. $y - 0.9 = 0.8y - 0.4$ _____

8. $3b + 4.7 = 6b - 7.9$ _____

9. $\frac{x}{4} + \frac{1}{3} = x - 8\frac{2}{3}$ _____

10. $\frac{n}{7} + 12 = n$ _____

11. $6x - 5 = 3x + 7$ _____

12. $\frac{x}{4} + 5 = \frac{x}{2} + 1$ _____

13. $2c - 7 = \frac{c}{2} + 5$ _____

14. $6y + 32 = 32y - 20$ _____

15. Beth has scores of 79, 85, 82, and 89 on her first four 100-point tests. What must she get on her fifth test to have an average of 85? _____

16. The cost, *C*, for baseballs is $2.95 per ball plus $3.54 tax. Jackson spent $92.04 on baseballs. How many baseballs did Jackson buy?

_____

Quick Warm-Up: Assessing Prior Knowledge
3.5 Using the Distributive Property

Solve for x.

1. $2 + 3x = 11$

2. $6x - 7 = 11$

3. $21 - 5x = 16$

_____

_____

_____

4. $\dfrac{x}{6} + 3 = 4$

5. $\dfrac{x}{9} - 3 = 6$

_____

_____

Lesson Quiz
3.5 Using the Distributive Property

Solve each equation.

1. $7x + 3 - 3x = 15$ _____

2. $4p + 1 - 5p = -8$ _____

3. $6(r + 2) - 4 = 20$ _____

4. $9 - 2(y - 3) = 11$ _____

5. $12t - 4(t - 5) = -4$ _____

6. $8x + 7(x + 2) = -1$ _____

7. $6(s - 4) + 1 = 3s + 7$ _____

8. $12 + 4n = 3n + 15$ _____

9. $-8 - 12q = -7 - 11q$ _____

10. $5(x + 3) = 7x + 5$ _____

11. $9w - 4(w - 2) = 3(w + 1) - 9$ _____

12. $7c + 2 - 4(2c + 1) = 3(c - 3)$ _____

One rectangle has sides whose lengths are $5x$ and $x + 10$. Another rectangle has sides whose lengths are $3x$ and $5x - 6$.

13. Write an equation to find the value of x for which the perimeters of

the two rectangles will be equal. _____

14. Solve your equation from Exercise 13. _____

Quick Warm-Up: Assessing Prior Knowledge
3.6 *Using Formulas and Literal Equations*

Solve for *x*.

1. $2(x + 3) = 6$

2. $3(x - 7) = 2x + 6$

3. $4(x + 3) = 5(2x + 1)$

_____

_____

_____

4. $7(2x - 4) = 14x - 28$

5. $6(2x - 5) = 12x + 2$

_____

_____

Lesson Quiz
3.6 *Using Formulas and Literal Equations*

**Convert each temperature from Fahrenheit to Celsius by using the
formula $C = \frac{5}{9}(F - 32)$.**

1. $14°F$ _____

2. $65°F$ _____

Solve each equation for the indicated variable.

3. $a + x = b$, for x _____

4. $c - d = f$, for c _____

5. $lw = A$, for l _____

6. $P - C = R$ for C _____

7. $rt = d$, for r _____

8. $2l + 2w = P$, for w _____

9. $y = 8x + b$, for b _____

10. $y = \frac{1}{3}x + b$, for x _____

11. $3r - 2s = 6$, for s _____

12. $8m - 4n = w$, for m _____

13. Use the formula for the area of a triangle, $A = \frac{1}{2}bh$, to find the height, h, if the area is 30

square inches and the base is 15 inches. _____

14. The formula for profit is $P = R - C$, where P stands for profit, R stands for revenue, and
C stands for cost. Use the formula to find how much revenue a company must generate to

make $150,000 in profit if the costs are $50,000. _____

Chapter Assessment

Chapter 3, Form A, page 1

Write the letter that best answers the question or completes the statement.

_____ 1. Solve $x + 0.54 = 2$.

    a. $x = 1.46$ **b.** $x = 2.54$ **c.** $x = 0.34$ **d.** $x = 0.46$

_____ 2. Solve $\frac{x}{9.2} = 8.1$.

    a. $x = 745.2$ **b.** $x = 74.52$ **c.** $x = 17.3$ **d.** $x = 73.52$

_____ 3. Solve $5x + 1.7 = 7x - 12.3$.

    a. $x = 7$ **b.** $x = 8$ **c.** $x = 5$ **d.** $x = 6$

_____ 4. Kendall wants to drive 280 miles in 7 hours. What speed in miles per hour should she average for the trip?

    a. 45   **b.** 40   **c.** 50   **d.** 55

_____ 5. Jerry pays \$1.40 per gallon for premium gasoline. He spends \$16.80 and wants to know how many gallons he bought. Which equation models this situation?

    a. $16.80 + x = 1.40$ **b.** $16.80 - 1.40 = x$

    c. $1.40 - x = 16.80$ **d.** $1.40x = 16.80$

_____ 6. Solve $3m - 17 = 19$.

    a. $m = 3$ **b.** $m = 36$ **c.** $m = 15$ **d.** $m = 12$

_____ 7. Solve $\frac{3x}{2} + 6 = 9$.

    a. $x = 10$ **b.** $x = -10$ **c.** $x = 2$ **d.** $x = -2$

_____ 8. Solve $6 - 6(x - 2) = 3(3x + 1)$.

    a. $x = 1$ **b.** $x = 0$ **c.** $x = -\frac{17}{15}$ **d.** $x = -\frac{1}{5}$

_____ 9. The cost of 10 pencils plus tax of \$0.15 is \$2.15. Which equation models the situation?

    a. $10(x + 0.15) = 2.15$ **b.** $10x + 0.15 = 2.15$

    c. $10x = 2.15 + 0.15$ **d.** $10 + 0.15x = 2.15$

Chapter Assessment

Chapter 3, Form A, page 2

_____ **10.** If $c - \frac{5}{8} = 6$, what is the value of c?

     a. $5\frac{3}{8}$ **b.** $-1\frac{7}{8}$ **c.** $6\frac{5}{8}$ **d.** $-5\frac{3}{8}$

_____ **11.** Solve $2x + 4 = 84$

     a. $x = -40$ **b.** $x = -37$ **c.** $x = 37$ **d.** $x = 40$

_____ **12.** Solve $2(9 - 2y) = 42$.

     a. $y = -6$ **b.** $y = 15$ **c.** $y = -15$ **d.** $y = 6$

_____ **13.** Solve $0.4r - 1.2 = 0.3r + 0.6$.

     a. $r = \frac{18}{7}$ **b.** $r = 6$ **c.** $r = 18$ **d.** not given

_____ **14.** Which equation has no solution?

     a. $2x + 4 = 2(x + 2)$ **b.** $2x + 4 = 2(x + 1)$

     c. $3x + 5 = 2(x + 3)$ **d.** Neither a nor b has a solution.

_____ **15.** Solve for P: $I = Prt$

     a. $P = I - rt$ **b.** $P = \frac{rt}{I}$ **c.** $P = \frac{I}{rt}$ **d.** $P = \frac{I \cdot r}{t}$

_____ **16.** The price of a shirt has been reduced by \$20. The cost of 10 shirts at the reduced price is \$50. Which equation models the situation?

     a. $50(x - 20) = 10$ **b.** $50(x + 20) = 10$

     c. $10(x + 20) = 50$ **d.** $10(x - 20) = 50$

_____ **17.** Solve for x: $y = mx + b$

     a. $x = \frac{y - b}{m}$ **b.** $x = \frac{b + y}{m}$ **c.** $x = \frac{y}{m} + b$ **d.** $x = y - \frac{b}{m}$

_____ **18.** A triangle has an area of 72 square feet. Use the formula $A = \frac{1}{2}bh$ to find the height of the triangle if the base is 24 feet.

     a. 60 feet   **b.** 6 feet   **c.** 864 feet   **d.** 1.5 feet

Chapter Assessment

Chapter 3, Form B, page 1

Write an equation or an inequality to represent each situation.

1. Paul has $10. Tom and Paul have $15 together. _____

   _____

2. Melissa has 5 times as much money as Manuel. They have $60 together. _____

   _____

3. The cost of 8 CD's plus a sales tax of $1.34 is $105.26. _____

   _____

Solve each equation.

4. $-4b = 48$ _____

5. $x + 9 = 23$ _____

6. $-3x = -1764$ _____

7. $\frac{x}{20} - 6 = 14$ _____

8. $\frac{c}{-3} = 0.3$ _____

9. $\frac{b}{-1} = -20$ _____

10. $5x - 2 = 13$ _____

11. $\frac{4}{5}y = 16$ _____

12. $\frac{-a}{5.2} = -7$ _____

13. $12x - (1 - 2x) = -29$ _____

14. $3x - 2(x + 3) = 4x - 7$ _____

15. $-3(x + 2) = -18$ _____

16. $2.9m + 1.7 = 3.5 + 2.3m$ _____

17. $\frac{3}{4}y - 4 = 7 + \frac{1}{2}y$ _____

18. Tara wants to buy one set of towels that costs $25.00 for the set and 6 decorative guest towels. Her total purchase was $40 before taxes. Write and solve an equation to determine how much she was

    charged for each decorative towel. _____

Chapter Assessment

Chapter 3, Form B, page 2

19. Solve $0.4x + 3.91 = 21.05$. _____

20. Solve $x = 0.5x + 3.8$. _____

21. Solve $5x + 2(3x - 1) = x$. _____

22. Solve $6x - 4(x + 1) = 2(x - 3) + 2$. _____

23. Solve $4(x + 1) = 2(x - 3) - (1 - 2x)$. _____

24. Convert 41°F to Celsius by using the formula $C = \frac{5}{9}(F - 32)$. _____

25. Solve $r = s - t$ for t. _____

26. Solve $ab - cd = 0$ for c. _____

27. The formula for the sum of the angles of the measures of a triangle is $a + b + c = 180$. Use the formula to find the measure of the third angle when the measures of two of the angles are 105° and 25°.

_____

28. A trapezoid has an area of 105 square inches. The shorter base measures 8 inches and the height is 10 inches. Use the formula, $A = \frac{1}{2}h(b_1 + b_2)$ to find the length of the longer base.

_____

29. The price of an automobile tire has been reduced by $15.00. The cost of a set of 4 tires at the reduced price is $600. Use an equation to find the original cost of a tire.

_____

Alternative Assessment

Exploring Addition and Subtraction Equations, Chapter 3, Form A

TASK: Solve problems by using addition and subtraction equations.

HOW YOU WILL BE SCORED: As you work through the task, your teacher will be looking for the following:

- whether you can assign a variable and write an equation for a situation
- how well you can explain the process of solving addition and subtraction equations
- whether you can use the solution of the equation to write an answer to the problem

Joni earns $50 per week more than Marshall. Billy's weekly salary is $20 less than Joni's weekly salary. If Marshall earns $320 per week, how much does Billy earn per week?

1. Explain how to identify the known and unknown information in the problem.

   _____

   _____

2. Describe how you would determine Joni's salary.

   _____

   _____

3. Discuss how you would set up an algebraic equation to describe Billy's weekly salary in terms of Joni's weekly salary.

   _____

   _____

4. Find Billy's weekly salary.

   _____

5. Describe two different ways to check your solution.

   _____

   _____

6. Use your solution to write an answer to the problem.

   _____

SELF-ASSESSMENT: Explain how you could have solved the problem by using guess-and-check.

Alternative Assessment
Exploring Multistep Equations, Chapter 3, Form B

TASK: Use multistep equations to make a decision.

HOW YOU WILL BE SCORED: As you work through the task, your teacher will be looking for the following:

- whether you can interpret the problem and decide what information you need in order to determine an answer
- whether you can write and solve an equation in order to find this information
- whether you can use your information to decide which company to use and can explain your decision

One Internet access company charges a flat fee of $25.00 per month. Another company charges a monthly fee of $15 plus a charge of $0.05 per minute for every minute of on-line time. Which company would you use?

1. Describe how the plans are set up differently.

_____

_____

2. What information do you need to find in order to make a decision?

_____

_____

3. Write and solve an equation in one variable in order to find the information described in Exercise 2.

_____

_____

4. When will the monthly charge for the two companies be the same?

_____

_____

5. Which company would you choose? _____

SELF-ASSESSMENT: Explain how algebra helped you evaluate the companies.

_____

_____

Quick Warm-Up: Assessing Prior Knowledge
4.1 *Using Proportional Reasoning*

Write as a fraction in simplest form.

1. $\frac{4}{16}$

2. $\frac{28}{35}$

3. $\frac{48}{64}$

4. $\frac{72}{44}$

_____   _____   _____   _____

Solve each equation.

5. $6k = 96$

6. $54 = 3n$

7. $32r = 4$

8. $0.5y = 15$

_____   _____   _____   _____

Lesson Quiz
4.1 *Using Proportional Reasoning*

Solve each proportion.

1. $\frac{5}{6} = \frac{15}{c}$ _____

2. $\frac{4}{5} = \frac{f}{20}$ _____

3. $\frac{9}{10} = \frac{h}{120}$ _____

4. $\frac{49}{56} = \frac{t}{8}$ _____

5. $\frac{y}{66} = \frac{10}{11}$ _____

6. $\frac{m}{49} = \frac{20}{7}$ _____

7. $\frac{3.8}{p} = \frac{2}{5}$ _____

8. $\frac{6}{2.5} = \frac{n}{5}$ _____

9. $\frac{z}{4.8} = \frac{3}{1.2}$ _____

10. The ratio $\frac{x + 16}{40}$ is equal to the ratio $\frac{1}{2}$. What is the value of x?

_____

11. The ratio of boys to girls in algebra class is 5 to 4. There are 16 girls in algebra class. How many boys are in the class?

_____

12. A 5-foot pole casts a shadow of 8 feet. How tall is a pole that will cast a shadow of 28 feet?

_____

Quick Warm-Up: Assessing Prior Knowledge
4.2 *Percent Problems*

Write as a decimal.

1. $\dfrac{4}{5}$

2. $1\dfrac{3}{8}$

3. $\dfrac{7}{100}$

4. $\dfrac{117}{100}$

_____ _____ _____ _____

Solve each equation.

5. $0.4y = 50$

6. $25t = 15$

7. $\dfrac{3}{100} = \dfrac{s}{75}$

8. $\dfrac{12}{100} = \dfrac{18}{a}$

_____ _____ _____ _____

Lesson Quiz
4.2 *Percent Problems*

Write each percent as a decimal.

1. 35% _____ 2. 12.5% _____ 3. 0.6% _____

Write each percent as a fraction or mixed number in lowest terms.

4. 75% _____ 5. 135% _____ 6. 0.4% _____

Find each answer.

7. Find 25% of 80. _____ 8. 20 is what percent of 60? _____

9. What percent of 40 is 10? _____ 10. Find 120% of 50. _____

11. 30 is what percent of 15? _____ 12. What percent of 150 is 15? _____

13. Find 12.5% of 64. _____ 14. 112.5 is what percent of 90? _____

15. A sweater originally costs $70. It is reduced by 30%. What is the sale price?

_____

16. A jacket is marked down from $75 to $60. By what percent is the jacket marked down?

_____

Algebra 1

Quick Warm-Up: Assessing Prior Knowledge
4.3 Introduction to Probability

Write each fraction as a decimal and as a percent.

1. $\frac{3}{4}$ _____

2. $\frac{7}{10}$ _____

3. $\frac{23}{100}$ _____

4. $\frac{1}{8}$ _____

5. $\frac{2}{9}$ _____

6. $\frac{125}{300}$ _____

Lesson Quiz
4.3 Introduction to Probability

Two coins were flipped 10 times with the following results:

Trial	1	2	3	4	5	6	7	8	9	10
Coin 1	T	T	H	H	H	H	T	H	H	T
Coin 2	H	H	H	T	H	T	T	H	H	H

Based on the results above, find the experimental probability of each outcome.

1. Both coins are alike. _____

2. Both coins are tails. _____

3. At least one coin is tails. _____

4. Neither coin is tails. _____

5. Keesha tossed a bottle cap 40 times. It landed up 23 times. Find the experimental probability that the bottle cap does not land up. _____

Two number cubes were rolled 50 times. Based on the results below, find the experimental probability of each outcome.

6. An even number appeared 35 times. _____

7. A sum greater than 6 appeared 28 times. _____

Two coins were flipped 150 times. Based on the results below, find the experimental probability of each outcome.

8. Two heads appeared 60 times. _____

9. One heads and one tails appeared 75 times. _____

Mid-Chapter Assessment

Chapter 4 (Lessons 4.1–4.3)

Write the letter that best answers the question or completes the statement.

_____ 1. If $\frac{a}{9} = \frac{6}{5}$, what is the value of a?

     a. 6   **b.** 10   **c.** 10.8   **d.** not given

_____ 2. What 80% of 75?

     a. 60   **b.** 600   **c.** 6   **d.** not given

_____ 3. What percent of 50 is 200?

     a. 25%   **b.** 200%   **c.** 400%   **d.** not given

_____ 4. A pair of number cubes is rolled 200 times. The sum 6 appears 90 times. What is the experimental probability of this outcome?

     a. 6%   **b.** 90%   **c.** 45%   **d.** not given

Solve each proportion.

5. $\frac{2}{3} = \frac{m}{9}$ _____   6. $\frac{5}{b} = \frac{25}{80}$ _____   7. $\frac{3.6}{1.2} = \frac{1.5}{t}$ _____

8. The ratio of algebra books to history books in the library is 3 to 5. There are 12 algebra books in the library. How many history books

    are in the library? _____

Find each answer.

9. Find 25% of 120. _____   10. What is 35% of 320? _____

11. 45 is what percent of 135? _____   12. 50 is what percent of 25? _____

13. Find 12.5% of 120. _____   14. What is 115% of 34? _____

15. The price of a CD increased from $12 to $15. What is the percent of increase? _____

16. A penny and a nickel were tossed 180 times. Heads appeared on both coins 108 times. What is the experimental probability of this outcome? _____

Quick Warm-Up: Assessing Prior Knowledge
4.4 *Measures of Central Tendency*

Arrange each set of numbers in order from least to greatest.

1. 7, 19, 1, 14, 4, 9, 15, 5

2. 1.1, 0.25, 7.8, 2.5, 11.1, 0.5

_____   _____

Find each difference.

3. 9.2 − 5.8 _____   4. 16 − 5.1 _____

Lesson Quiz
4.4 *Measures of Central Tendency*

Find the mean, median, mode, and range for each set of data.

1. 13, 12, 13, 8, 10, 5, 14, 15, 11 _____

2. 80, 82, 82, 79, 90, 95, 80, 82, 90, 100 _____

Day	Temperature
Monday	88°F
Tuesday	90°F
Wednesday	92°F
Thursday	88°F
Friday	95°F
Saturday	90°F
Sunday	90°F

The high temperatures in Denver for a week during August are shown in the table.

3. Find the mean, median, mode, and range of the temperatures.

_____

4. If the high temperature on Friday had been 70°, would this change the mean, median, mode, or range? Explain.

_____

Number of hours per day	1	2	3	4
Frequency	ʍ	ʍ ʍ ‖	ʍ ‖‖‖‖	ʍ ‖

A survey was conducted to determine the number of hours of homework high school students do per day. The data is shown in the frequency table.

5. How many students responded to the survey? _____

6. What are the mode and mean of the data? _____

Quick Warm-Up: Assessing Prior Knowledge
4.5 *Graphing Data*

Write as a decimal. Round to the nearest thousandth if necessary.

1. 68%

2. 3.2%

3. $\frac{37}{50}$

4. $\frac{37}{150}$

_____ _____ _____ _____

Evaluate each expression.

5. $0.21 \times 40{,}000$

6. 42% of 7300

_____ _____

Lesson Quiz
4.5 *Graphing Data*

1. If you took a survey at school to find the four most popular sodas, which type of graph would you use to display your results and why?

_____

Use the bar graph for Exercises 2–3.

2. Estimate the difference in the amount of rental sales for the film with the highest rental sales and the film with the lowest rental sales.

_____

3. Which two movies have rental sales that are within 10 million dollars of each other?

_____

Use the circle graph for Exercises 4–5.

4. If the company's total budget is $360,000, how much

does it spend on marketing? _____

5. If your department received less than $62,000, in which

department do you work? _____

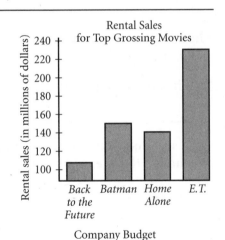

Quick Warm-Up: Assessing Prior Knowledge
4.6 *Other Data Displays*

1. Find the mean, median, mode(s), and range for each set of data.

Number	Frequency
11	||||
12	||
16	||||
19	|
22	||

_____

_____

Lesson Quiz
4.6 *Other Data Displays*

Each number in the table represents the average speed, in miles per hour, of the winning car in 26 runs of the Indianapolis 500.

139	139	140	143	147	151	144
151	153	157	156	158	163	159
159	149	149	161	161	159	143
139	162	162	164	159		

1. Construct a stem-and-leaf plot of the data in the table above.

2. What are the mean, median, and mode of the data? _____

3. What is the most common value in the data set? What measure of central tendency do you think best answers the question? Why?

_____

4. Use the stem-and-leaf plot you made in Exercise 1 to make a box-and-whisker plot.

_____

5. If a winning car's average speed in another run of the Indianapolis 500 is 157 miles per hour, where would it

fall on the box-and-whisker plot? _____

6. How would you construct a histogram for the data? _____

_____

Chapter Assessment

Chapter 4, Form A, page 1

Write the letter that best answers the question or completes the statement.

_____ 1. Which of the following is a true proportion?

 a. $\frac{7}{8} = \frac{8}{7}$ b. $\frac{16}{28} = \frac{36}{63}$ c. $\frac{4}{7} = \frac{12}{28}$ d. $\frac{18}{45} = \frac{3}{15}$

_____ 2. Solve the proportion $\frac{5}{8} = \frac{15}{c}$.

 a. 45   b. 3   c. 24   d. 16

_____ 3. What is 3.687 written as a percent?

 a. 368.7%   b. 0.03687%   b. 36.76%   d. 0.3687%

_____ 4. What is 37% of 50?

 a. 0.185   b. 18.5   b. 185   d. 1.85

_____ 5. 15 is what percent of 45?

 a. $66\frac{2}{3}\%$ b. 30%   b. 300%   d. $33\frac{1}{3}\%$

_____ 6. 40% of what number is 20?

 a. 0.05   b. 0.5   b. 5   d. 50

_____ 7. The cost of a jacket is increased from $75 to $80. What is the percent of increase?

 a. 87.5%   b. $6\frac{2}{3}\%$ b. 6.25%   d. $66\frac{2}{3}\%$

A number cube is rolled 10 times. Use the table of results to find each probability.

Trial	1	2	3	4	5	6	7	8	9	10
Number rolled	4	3	6	6	1	5	2	6	2	5

_____ 8. rolling 1 on the number cube

 a. 10%   b. 20%   c. 40%   d. 50%

_____ 9. rolling 5 on the number cube

 a. 10%   b. 20%   c. 40%   d. 50%

Chapter Assessment

Chapter 4, Form A, page 2

Write the letter that best answers each question or completes each statement.

The bar graph shows the percent of 18-year-olds with high school diplomas for the given years.

_____ 10. What year had the highest percent of 18-year-olds with high school diplomas?

    a. 1975   **b.** 1970

    c. 1985   **d.** not given

_____ 11. Between what consecutive years did the greatest increase or decrease in the percent of 18-year-olds with diplomas occur?

    a. 1955 and 1960   **b.** 1970 and 1975

    c. 1950 and 1955   **d.** not given

_____ 12. According to the circle graph, what items are you likely to have the most of in your shopping cart?

    a. meat   **b.** produce

    c. canned goods   **d.** not given

_____ 13. If you have 50 items in your cart, how many would probably be seafood items?

    a. 5   **b.** 10

    c. 20   **d.** none

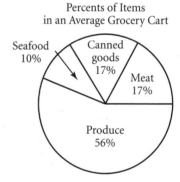

Percents of Items in an Average Grocery Cart

Use the frequency table to answer Exercises 14–16.

_____ 14. What is the mean of the data?

    a. 80.4   **b.** 80

    c. 82   **d.** not given

_____ 15. What is the mode of the data?

    a. 78   **b.** 89

    c. 95   **d.** not given

_____ 16. What is the median of the data?

    a. 79   **b.** 80.4

    c. 80   **d.** not given

Grades	60	64	72	78	80	89	95
Frequency	II	III	II	ҬҤҬ	ҬҤҬ III	IIII	ҬҤҬ I

Chapter Assessment

Chapter 4, Form B, page 1

Solve each proportion.

1. $\dfrac{4}{5} = \dfrac{8}{p}$ _____

2. $\dfrac{q}{16} = \dfrac{3}{4}$ _____

3. $\dfrac{6}{r} = \dfrac{15}{40}$ _____

4. Roberto earned $6.50 by mowing lawns in his neighborhood for

    2 hours. How much will he earn by mowing lawns for 5 hours? _____

Write each percent as a decimal.

5. 72% _____

6. 7.5% _____

7. 112% _____

Write each percent as a fraction in lowest terms.

8. 4.5% _____

9. 20% _____

10. 104% _____

Find each answer.

11. What is 35% of 200? _____

12. 40 is what percent of 25? _____

13. 25 is what percent of 125? _____

14. Find 115% of 80. _____

15. What number is 70% of 15? _____

16. What is 2.5% of 120? _____

17. The sales tax is 7.5%. If you buy $80 worth of merchandise, how

    much is your total bill? _____

Two number cubes were rolled 10 times. Use the table of results to find the experimental probability of each outcome.

Trial	1	2	3	4	5	6	7	8	9	10
Cube 1	6	5	4	1	1	5	3	6	4	6
Cube 2	4	3	4	3	2	5	2	2	3	1

18. 5 is rolled on the first cube. _____

19. 2 is rolled on the second cube. _____

20. Doubles are rolled. _____

21. Even numbers are rolled on both cubes. _____

Chapter Assessment

Chapter 4, Form B, page 2

The line graph shows the typing speed of individuals who practice different amounts of time during the week.

22. What is the typing speed of the person who practices 9 hours per week? _____

23. How many hours did a person practice to achieve a typing speed of 75 words per minute (wpm)? _____

24. What is the increase in typing speed from a person who practices 2 hours a day to a person who practices 11 hours a day? _____

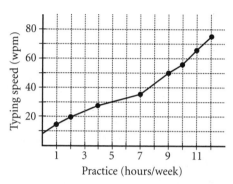

25. What does this graph tell you about typing speed in terms of hours of practice?

_____

26. Find the mean, median, mode, and range for the following data:

$$21, 19, 18, 17, 22, 44$$

_____

27. If another value of 18 were added to the list of numbers above, how would it affect the measures of central tendency?

_____

28. Create a stem-and-leaf plot for the following data:

14	20	31	40	55	16
27	35	16	33	27	27
26	32	24	23	46	21

29. Using the stem-and-leaf plot from Exercise 28, find the mean, median, mode, and range of the data.

_____

30. Find the upper and lower quartiles for the stem-and-leaf plot from Exercise 28.

_____

31. Using the results from Exercises 29 and 30, make a box-and-whisker plot.

_____

Alternative Assessment

Exploring Proportion Problems, Chapter 4, Form A

TASK: Use proportions to solve problems.

HOW YOU WILL BE SCORED: As you work through the task, your teacher will be looking for the following:
- whether you can make a table of values for the points shown in a graph
- how well you can write an equation to describe the values in a table
- a method for determining whether the equation represents a proportion

This graph shows the relationship between the number of tablespoons of a liquid plant fertilizer and the number of gallons of water.

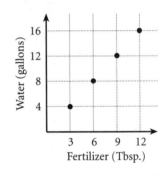

1. Complete the table by using the data from the graph.

Number of tablespoons of fertilizer	3	6	9	12
Number of gallons of water				

2. Write a sentence to describe the ratio of the number of tablespoons of fertilizer to the number of gallons of water.

_____

3. Discuss how you would write an equation to find the number of gallons of water to add to 9 tablespoons of fertilizer. Then write and solve the equation.

_____

_____

4. Describe two different methods for determining if your equation is a proportion.

_____

_____

5. How many tablespoons of fertilizer are required for 16 gallons of water?

_____

6. How many gallons of water are required for 2 tablespoons of fertilizer?

_____

SELF-ASSESSMENT: Explain the difference between a ratio and a proportion.

Alternative Assessment
Graphing Data, Chapter 4, Form B

TASK: Analyze a set of data by drawing graphs.

HOW YOU WILL BE SCORED: As you work through the task, your teacher will be looking for the following:

- whether you can evaluate the effect of using different scales to graph data
- whether you can describe the type of information displayed by a box-and-whisker plot

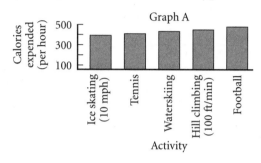

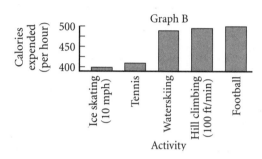

1. Using graph A, estimate the Calories expended per hour for ice skating, tennis, waterskiing, hill climbing, and football. Then estimate by using graph B.

_____

_____

2. Which graph is the best graph for displaying the data? Explain why.

_____

Ed bicycled to the bus stop, parked his bike, and rode the bus to the beach. At the beach, he played volleyball, went for a swim, and relaxed on the beach. The data shows the amount of time and the number of calories expended for each activity.

Activity	Time (in minutes)	Calories expended
Bicycling	20	70
Sitting	45	75
Volleyball	60	350
Swimming	30	150
Lying down	45	40

The box-and-whisker plot displays the total number of calories expended for each activity.

3. What type of information is best displayed by the

   box-and-whisker plot? _____

Total Number of Calories Expended

40   75   250   350

SELF-ASSESSMENT: Choose a typical day and record the amount of time you spend on different activities. Display the information in a bar graph.

Quick Warm-Up: Assessing Prior Knowledge
5.1 *Linear Functions and Graphs*

Graph each point on a coordinate plane.

1. $A(3, 3)$

2. $B(6, 2)$

3. $C(1, 0)$

4. $D(0, 5)$

5. $E(7, 4)$

6. $F(4, 7)$

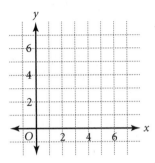

Lesson Quiz
5.1 *Linear Functions and Graphs*

Determine whether the relation is a function.

1. $\{(2, 3), (-3, 9), (2, 5), (5, 2)\}$ _____

2. $\{(7, 13), (8, 4), (-12, 4)\}$ _____

3. $\{(8, 24), (6, 24), (2, -24), (3, 2)\}$ _____

4. $\{(4, 3), (-5, 12), (19, 6)\}$ _____

Complete each ordered pair so that it is a solution to $-5x + y = 7$.

5. $(0, ?)$ _____

6. $(6, ?)$ _____

7. $(13, ?)$ _____

8. $(-3, ?)$ _____

9. $(-10, ?)$ _____

10. $(3, ?)$ _____

11. $(?, 17)$ _____

12. $(?, 27)$ _____

13. $(?, -3)$ _____

Quick Warm-Up: Assessing Prior Knowledge
5.2 *Defining Slope*

Evaluate.

1. $4 - 9$ _____

2. $6 - (-4)$ _____

3. $\dfrac{-8 - 4}{9 - 5}$ _____

4. $\dfrac{-5 - (-5)}{1 - (-1)}$ _____

Lesson Quiz
5.2 *Defining Slope*

Find the slope of the line that contains each pair of points:

1. A and B _____

2. C and D _____

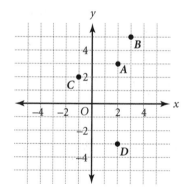

Find the slope for each rise and run.

3. rise: 4, run: 6 _____

4. rise: -6, run: 8 _____

5. rise: 7, run: -2 _____

6. rise: -1.9, run: -3.8 _____

Find the slope of the line that contains each pair of points.

7. $M(4, 1), N(2, 3)$ _____

8. $M(-2, 0), N(3, -4)$ _____

9. $M(-1, -3), N(4, 1)$ _____

10. $M(7, 1), N(-3, 1)$ _____

11. $M(-8, 2), N(-4, -4)$ _____

12. $M(1.5, -4.5), N(-3, -4)$ _____

Quick Warm-Up: Assessing Prior Knowledge
5.3 *Rate of Change and Direct Variation*

Solve each equation.

1. $15t = 3$ _____

2. $42 = 0.6m$ _____

3. $(12)(9) = d$ _____

4. $r = (8.4)(1.5)$ _____

Lesson Quiz
5.3 *Rate of Change and Direct Variation*

In a thunderstorm, lightning is seen as it occurs, but the sound of thunder is heard a short time later. The graph at right shows the distance that thunder travels as a function of time.

1. How far does the thunder travel in 5 seconds? _____

2. Find the speed of thunder. _____

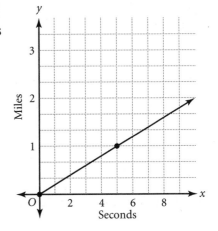

Andrew drove away from the grocery store at a constant speed. The graph at right shows his distance from his apartment in terms of the number of minutes since he left the store.

3. At what speed was Andrew traveling before he stopped?

   _____

4. How long was he stopped?

   _____

5. How long did it take him to get home?

   _____

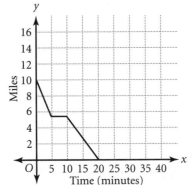

Mid-Chapter Assessment

Chapter 5 (Lessons 5.1–5.3)

Write the letter that best answers the question or completes the statement.

_____ 1. What is the slope of a line that has a rise of -3 and a run of 9?

     a. 3   **b.** -3 **c.** $\frac{1}{3}$ **d.** $-\frac{1}{3}$

_____ 2. Find the slope of the line that contains $M(3, 4)$ and $N(-1, 2)$.

     a. $-\frac{3}{2}$ **b.** $\frac{1}{2}$ **c.** 2   **d.** $-\frac{1}{2}$

_____ 3. In the direct-variation equation $y = 2.4x$,

     a. x is the constant of variation.
     b. y is the rate of change.
     c. the rate of change is the reciprocal of 2.4.
     d. the rate of change is 2.4.

_____ 4. Let $y = kx$ be an equation for a direct variation. Which statement is *not* true?

     a. The constant of variation is the same as the slope of the graph of $y = kx$.
     b. The constant of variation is k.
     c. The graph of a direct-variation equation does not pass through the origin.
     d. The rate of change is the ratio $\dfrac{\text{change in } y}{\text{corresponding change in } x}$.

Find the slope of the line with each rise and run.

5. rise: 8; run: 4 _____   6. rise: 15; run: -20 _____

Find the slope of the line that contains each pair of points.

7. $M(-3, 8), N(2, 5)$ _____   8. $M(4, -5), N(-4, -5)$ _____

9. $M(-6, 4), N(-2, 3)$ _____   10. $M(1, 1.2), N(0, 3.6)$ _____

For the following values of *x* and *y*, *y* varies directly as *x*. Find the constant of variation, and write an equation of direct variation.

11. $y = 2$ when $x = 5$ _____

12. $y = 7$ when $x = 2.5$ _____

13. $y = 1.5$ when $x = 3$ _____

14. $y = 7.6$ when $x = 5.7$ _____

Quick Warm-Up: Assessing Prior Knowledge
5.4 The Slope-Intercept Form

Find the slope of the line that contains each pair of points.

1. $A(3, 6)$, $B(1, 12)$ _____

2. $C(0, -2)$, $D(-6, -2)$ _____

Complete each ordered pair so that it is a solution of the equation $y = -2x + 3$.

3. $(2, \underline{?})$ _____

4. $(0, \underline{?})$ _____

5. $(\underline{?}, 11)$ _____

6. $(\underline{?}, 0)$ _____

Lesson Quiz
5.4 The Slope-Intercept Form

Match each equation with the appropriate description.

1. $x = 5$ _____

2. $y = 5$ _____

3. $x = 5y$ _____

a. a horizontal line 5 units above the origin

b. a line through the origin with a slope of $\frac{1}{5}$

c. a vertical line 5 units to the right of the origin

Find the slope and y-intercept for the graph of each equation.

4. $y = -5x + 3$ _____

5. $y = \frac{2}{3}x - \frac{7}{3}$ _____

Write an equation in slope-intercept form for the line that fits each description below.

6. slope of -3 and y-intercept of 5 _____

7. contains $(0, 7)$ and has a slope of -7 _____

Write an equation for the line that contains each pair of points.

8. $(-1, 7)$, $(0, 3)$ _____

9. $(-7, 2)$, $(5, -1)$ _____

10. $(4, 3)$, $(-2, -9)$ _____

11. $(3, -7)$, $(-1, -3)$ _____

Quick Warm-Up: Assessing Prior Knowledge
5.5 The Standard and Point-Slope Forms

Solve for the indicated variable.

1. $t = r + s$ for s _____

2. $-5g = h$ for g _____

3. $q = mn - p$ for m _____

4. $4x + 2y = z$ for y _____

Lesson Quiz
5.5 The Standard and Point-Slope Forms

Write each equation in standard form.

1. $3y = 9x + 6$ _____

2. $5y = 3x - 10$ _____

3. $5x - 4 = 3y + 2$ _____

4. $4 + \frac{1}{2}y = x$ _____

Write an equation in point-slope form for the line with each slope that contains the given point.

5. slope: 3; $(1, 2)$ _____

6. slope: -1; $(3, 2)$ _____

7. slope: $\frac{1}{2}$; $(-2, 7)$ _____

8. slope: $-\frac{3}{2}$; $(3, 0)$ _____

Find the x- and y-intercepts for the graph of each equation.

9. $x - y = 5$ _____

10. $2x + 3y = 6$ _____

11. $2x - 5y = -20$ _____

12. $3x - y = 3.6$ _____

Write an equation in standard form for the line that contains each pair of points.

13. $(2, 1), (-1, 3)$ _____

14. $(0, 1), (-1, 1)$ _____

Write an equation in standard form for each line described below.

15. contains the point $(2, -3)$ and has a slope of 5 _____

16. crosses the x-axis at $x = -3$ and the y-axis at $y = 1$ _____

Quick Warm-Up: Assessing Prior Knowledge
5.6 *Parallel and Perpendicular Lines*

Find each product.

1. $(3)\left(\frac{1}{3}\right)$ _____

2. $\left(\frac{4}{7}\right)\left(-\frac{7}{4}\right)$ _____

Identify the slope and the *y*-intercept of each line. Solve for *y* if necessary.

3. $y = -2x + 5$ _____

4. $2x + 2y = 5$ _____

Lesson Quiz
5.6 *Parallel and Perpendicular Lines*

Graph the line $y = 2x$ on the grid provided. On the same grid,

1. write the equation for a line that is parallel to $y = 2x$. Then graph the line.

   _____

2. write the equation for a line that is perpendicular to $y = 2x$. Then graph the line.

   _____

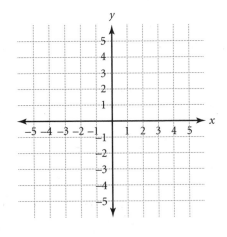

Write an equation in slope-intercept form for the line that contains the point (2, −1) and is

3. parallel to the line $3x - 6y = 12$. _____

4. perpendicular to the line $3x - 6y = 12$. _____

Write an equation in slope-intercept form for the line that contains the point (1, 2) and is

5. parallel to the line $y = -x + 5$. _____

6. perpendicular to the line $y = -x + 5$. _____

Chapter Assessment

Chapter 5, Form A, page 1

Write the letter that best answers the question or completes the statement.

_____ 1. What is the slope of a line that passes through the origin and the point $(-2, 1)$?

    a. $-\dfrac{1}{2}$ b. 2   c. $\dfrac{1}{2}$ d. -2

_____ 2. What is the slope of a line with a rise and a run of 4?

    a. $\dfrac{4}{5}$ b. $-\dfrac{4}{5}$ c. $\dfrac{5}{4}$ d. 1

_____ 3. An equation of a line that has a slope of 0 and contains the point $(0, -2)$ is

    a. $x = 2$.   b. $x = -2$.   c. $y = -2$.   d. $y = x - 2$.

_____ 4. In a graph of a linear function, the change in y is 3. The change is x is 9. The rate of change is

    a. 3.   b. 9.   c. $\dfrac{1}{3}$.   d. $\dfrac{1}{9}$.

_____ 5. The slope of the line containing the points $(2, -4)$ and $(3, -7)$ is

    a. -11.   b. -3.   c. $-\dfrac{1}{3}$.   d. 3.

_____ 6. Write the equation $y = \dfrac{1}{2}x + 5$ in standard form.

    a. $x - 2y = 10$ b. $x - 2y - 10 = 0$ c. $-x + 2y = 10$ d. $x - 2y = -10$

_____ 7. Which line is parallel to $y = \dfrac{1}{4}x + 2$?

    a. $y = \dfrac{1}{4}x + 5$ b. $y = -4x + 5$ c. $y = -\dfrac{1}{4}x + 5$ d. $y = 4x + 5$

_____ 8. If y varies directly as x and $y = 8$ when $x = 3$, find an equation of direct variation.

    a. $y = \dfrac{3}{8}x$ b. $y = -\dfrac{8}{3}x$ c. $y = \dfrac{8}{3}x$ d. $y = 8x$

Chapter Assessment

Chapter 5, Form A, page 2

_____ 9. If y varies directly as x and $y = \frac{2}{3}$ when $x = \frac{1}{2}$, find an equation of direct variation.

    a. $y = \frac{3}{4}x$ **b.** $y = -\frac{4}{3}x$ **c.** $y = \frac{4}{3}x$ **d.** $y = -\frac{3}{4}x$

_____ 10. What are the slope and y-intercept of the equation $3x - 4y = 4$?

    a. slope: -3; y-intercept: -4 **b.** slope: $\frac{3}{4}$; y-intercept: -1

    c. slope: 3; y-intercept: 4 **d.** slope: $\frac{3}{4}$; y-intercept: -4

_____ 11. What is the slope of a line perpendicular to $x - 3y = 4$?

    a. -3 **b.** 3 **c.** $\frac{1}{3}$ **d.** $-\frac{1}{3}$

_____ 12. Which line contains the point $(0, 2)$ and is perpendicular to $y = \frac{1}{4}x + 2$?

    a. $y = \frac{1}{4}x + 1$ **b.** $y = -\frac{1}{4}x + 2$ **c.** $y = -4x + 2$ **d.** $y = 4x + 2$

_____ 13. Which point is on the graph of the equation $3x + 6y = 27$?

    a. $(0, 9)$ **b.** $(5, 2)$ **c.** $(1, -5)$ **d.** $(-5, 7)$

_____ 14. An equation for the line that contains $(2, 4)$ and has an undefined slope is

    a. $y = 2x + 4.$ **b.** $x = 2.$ **c.** $y = 2.$ **d.** $y = 0.$

_____ 15. An equation for the line that crosses the x-axis at $x = -1$ and the y-axis at $y = 3$ is

    a. $3x + y = 3.$ **b.** $3x = y + 3.$ **c.** $y = 3x + 3.$ **d.** $y + 3x = -3.$

_____ 16. Which equation represents a horizontal line?

    a. $y = x$ **b.** $y = -2$ **c.** $x = 0$ **d.** $y = 2x + 1$

Chapter Assessment
Chapter 5, Form B, page 1

Each pair of points is on a line. Find the slope of each line.

1. $M(0, 1), N(1, 1)$ _____

2. $M\left(4, 2\frac{1}{2}\right), N(0, 2)$ _____

Find an equation of a line that contains the origin and the given point.

3. $(2, -2)$ _____

4. $(3, 4)$ _____

Suppose that y varies directly as x. Write an equation for the direct variation.

5. when $x = 2, y = 8$ _____

6. $y = 3.6$ when $x = 6$ _____

Find the y-intercept of the graph of each equation.

7. $y = -3x$ _____

8. $3x - 6y = 12$ _____

Write an equation for the specified line.

9. slope: 4; y-intercept: -3 _____

10. contains $(2, -9)$; slope: -1 _____

11. contains $(3, -1)$ and $(6, -3)$ _____

12. contains $(6, -12)$; slope: -2 _____

Marta worked 6 hours and earned \$51.

13. Write a direct-variation equation that models the situation. _____

14. At the same rate, how much will Marta earn in 9 hours? _____

15. Write the equation $y + 1 = \frac{1}{2}(x - 6)$ in standard form. _____

16. Write the equation $8x + 4y = 24$ in slope-intercept form. _____

17. Robert ran 6 miles in 72 minutes. What was his speed in miles per hour? _____

Write an equation for each line described below. Write your answer in slope-intercept form.

18. crosses the x-axis at $x = -2$ and the y-axis at $y = -1$ _____

19. contains the point $(2, 4)$ and is perpendicular to the graph of $y = \frac{1}{3}x + 1$ _____

20. contains the point $(-1, 0)$ and is parallel to the graph of $2x + 3y = 6$ _____

Chapter Assessment

Chapter 5, Form B, page 2

21. Graph the line $y = 3x$. Then sketch the graph of a line that is parallel to $y = 3x$ and write its equation.

22. Graph the line $y = \frac{3}{4}x + 2$. Then sketch the graph of a line that is perpendicular to $y = \frac{3}{4}x + 2$ and write its equation.

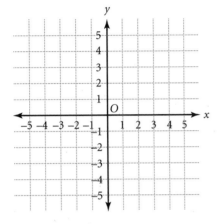

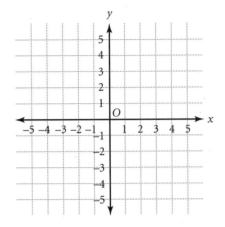

23. Draw the graph of the line with an x-intercept of $\frac{4}{3}$ and y-intercept of $\frac{4}{3}$.

24. Graph the equation $x - 4y = 2$ by finding the intercepts.

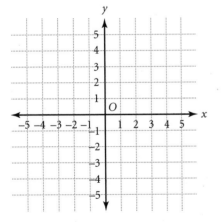

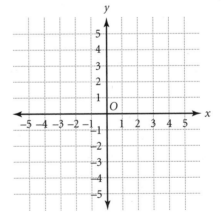

NAME _____ CLASS _____ DATE _____

Alternative Assessment

On the Slopes, Chapter 5, Form A

TASK: Use pictorial models to connect the concept of slope and steepness to real-world situations.

HOW YOU WILL BE GRADED: As you work through the task, your teacher will be looking for the following:

- whether you can find the slope of a line from its graph
- whether you can relate steepness to slope
- whether you can write the equation of a line from its graph

Two skiers leave from the same point and follow different cross-country trails.

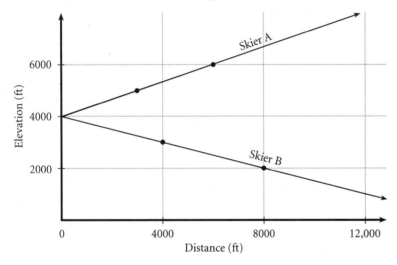

1. Calculate the slope of the line for each skier.

   skier A _____   skier B _____

2. Compare the slopes of the lines.

   _____

3. How does the steepness of a line compare to its slope?

   _____

4. Write the equation of the line for each skier in *y*-intercept form.

   skier A _____   skier B _____

5. Write a problem of your own that includes graphs of lines.

   _____

SELF-ASSESSMENT: Explain how the concepts of slope and steepness are related. Describe what you learned about this relationship from this worksheet.

Alternative Assessment

Sports Car Rally, Chapter 5, Form B

TASK: Interpret and create distance-time and speed-time graphs, and use the slopes of these graphs to draw conclusions about the speeds and distances traveled.

HOW YOU WILL BE SCORED: As you work through the task, your teacher will be looking for the following:

- how well you can determine the *y*-intercept on a graph
- whether you can describe the relationship between lines on a graph
- how well you can draw conclusions from graphs

The Bandameer Speedway is having a sports car rally. The graph shows the distance and time for the two cars in the first race.

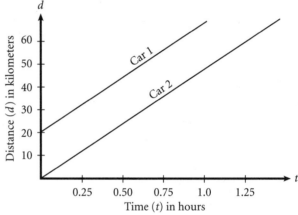

1. How fast is car 1 going? _____

2. How fast is car 2 going? _____

3. What do the parallel lines suggest?

_____

4. What is the *d*-intercept of the graph for car 1?

_____

5. What do the *d*-intercepts indicate about the two cars? _____

6. Write the equation of the line for car 1 in point-slope form. Write the equation of the line for car 2 in standard form.

_____

Suppose that car 1 is driven for 3 hours at the speed found in Step 1.

7. Draw a graph to show the relationship between speed and time.

8. Describe the graph of the line and write its equation.

_____

9. Draw a perpendicular line on the graph that intersects the time axis at 3 hours. What is the equation of this line?

_____

SELF-ASSESSMENT: Explain how this activity shows the relationship between equations of a line and time, distance, and speed.

Quick Warm-Up: Assessing Prior Knowledge
6.1 Solving Inequalities

Write the number that corresponds to each lettered point on this number line.

1. A _____

2. B _____

3. C _____

4. D _____

5. E _____

6. F _____

Lesson Quiz
6.1 Solving Inequalities

Solve each inequality.

1. $x - 9 < 5$ _____

2. $x - 5 \leq 8$ _____

3. $x + \frac{2}{3} < 1$ _____

4. $x + \frac{1}{2} \geq \frac{5}{6}$ _____

Solve each inequality and graph the solution on a number line.

5. $-7 + t \leq 3$

6. $u - 5 > -2.5$

7. $\frac{4}{8} < v + \frac{3}{8}$

8. $-3.5 \geq 2.5 + m$

Write an inequality that describes the set of points graphed on each number line.

9.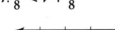

10.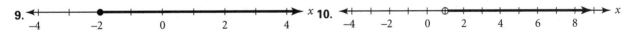

_____

_____

Write an inequality to represent the statement below.

11. Bowlers receive a second place trophy if their team average is at

least 180 but not more than 199. _____

Quick Warm-Up: Assessing Prior Knowledge
6.2 *Multistep Inequalities*

Solve each equation. Check your solution.

1. $-1.6t = -5$

_____

2. $b - 3.4 = -6$

_____

3. $4a + 6 = 9 + 8a$

_____

4. $\frac{z}{3} - 3 = -12$

_____

Lesson Quiz
6.2 *Multistep Inequalities*

Solve each inequality.

1. $\frac{x}{-4} \geq -5$ _____

2. $\frac{7f}{-2} \leq 14$ _____

3. $2x - \frac{3}{5} > \frac{4}{10}$ _____

4. $\frac{-h}{16} < \frac{1}{4}$ _____

5. $x - 7 < 2$ _____

6. $-5 > c + (-2)$ _____

7. $9 - x \leq 3$ _____

8. $7x > 5$ _____

9. $10 + \frac{x}{5} < 20$ _____

10. $\frac{y}{6} + 5 > 6$ _____

11. $2x + 5 > -9 + 3x$ _____

12. $-2y + 1 < 7 - 6y$ _____

Sal has \$125 to buy T-shirts for his softball team. Each shirt costs \$17.99 and there is a one-time charge of \$15 to print the team name on the shirts.

13. Write an inequality that expresses the maximum number of T-shirts

he can buy. _____

14. What is the maximum number of shirts he can buy? _____

Quick Warm-Up: Assessing Prior Knowledge
6.3 *Compound Inequalities*

Solve each inequality.

1. $16 + h < 25$

2. $4m \geq 22$

Solve each inequality and graph.

3. $6x - 10 \geq 14$

4. $15 - 2y > 21$

Lesson Quiz
6.3 *Compound Inequalities*

Graph each solution set.

1. $-1 < x < 4$

2. $x < 3$ or $x > 4$

3. $0 \leq y < 7.5$

4. $k < 2$ or $k \geq 3.5$

5. $2.0 < m \leq 5.5$

6. $d > 15$ or $d \leq -17$

7. $x \leq -23$ or $x \geq -4$

8. $d \geq 2$ or $d < -7$

9. $3.25 \leq y < 20$

Solve the following compound inequalities.

10. $-5 \leq x + 2 < 0$ _____

11. $10p > -2$ or $2 + p \leq -15$ _____

12. $-3 < 3j - 4 < 5$ _____

13. $-4t < 8$ or $-3t > 12$ _____

14. $\frac{-1}{2} < u + 1 \leq 2$ _____

15. $-4 < 5j - 3 < 7$ _____

Mid-Chapter Assessment

Chapter 6 (Lessons 6.1–6.3)

Write the letter that best answers the question or completes the statement.

_____ 1. Solve $x + 8 > -2$.

         a. $x > 6$ b. $x < 6$ c. $x > -10$ d. $x < -10$

_____ 2. Solve $2x + 1 < 5$.

         a. $x < 2$ b. $x > 2$ c. $x < 4$ d. $x > 4$

_____ 3. Solve $-5y > 15$.

         a. $x > -3$ b. $y > 20$ c. $y < 3$ d. $y < -3$

_____ 4. Solve $6 - x \leq 2$.

         a. $x \geq 4$ b. $x \geq -4$ c. $x \leq 4$ d. $x \leq -4$

_____ 5. Solve $9p + 7 \geq 9 + 10p$.

         a. $p \leq 2$ b. $p \geq 2$ c. $p \leq -2$ d. $p \geq -2$

_____ 6. Solve $\dfrac{-y}{6} + 4 < 1$.

         a. $y < -\dfrac{1}{2}$ b. $y > 18$ c. $y < 18$ d. $y < -18$

7. Write an inequality that describes the set of points graphed on the number line.

8. The maximum occupancy of an elevator is 15 people. There are 8 people in the elevator when it stops at the tenth floor where it becomes an express elevator with no more stops. Write an inequality to describe the maximum number of people who can get on the elevator on the tenth floor.

The school band is raising money to help buy new uniforms. They need to raise at least $2000. How many $16 concert tickets do they need to sell?

9. Write an inequality to model this situation.

10. How many concert tickets do they need to sell to reach their goal?

NAME _____ CLASS _____ DATE _____

Quick Warm-Up: Assessing Prior Knowledge
6.4 Absolute-Value Functions

Evaluate.

1. $|-11|$ _____ **2.** $|2-6|$ _____ **3.** $|6-2|$ _____

Find the distance between each pair of points on a number line.

4. $1, 7$ _____ **5.** $-2, 3$ _____

Lesson Quiz
6.4 Absolute-Value Functions

Evaluate.

1. $|7|$ _____ **2.** $|-12|$ _____

3. $|4-10|$ _____ **4.** $|10-4|$ _____

5. $|-10-10|$ _____ **6.** $|-10-(-10)|$ _____

Find the domain and range of each function.

7. $y = |x|$ Domain _____ Range _____

8. $y = -2|x|$ Domain _____ Range _____

9. $y = |x| + 3$ Domain _____ Range _____

10. $y = |x - 3|$ Domain _____ Range _____

11. Graph the function $y = |x - 3|$.

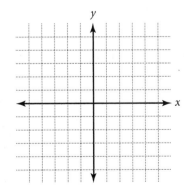

Copyright © by Holt, Rinehart and Winston. All rights reserved.

Holt Algebra 1 **Assessment Resources** **75**

Quick Warm-Up: Assessing Prior Knowledge
6.5 *Absolute-Value Equations and Inequalities*

Evaluate each expression.

1. $|-6|$
2. $-|-2|$
3. $-|5.1|$
4. $|0|$

_____   _____   _____   _____

Simplify each expression.

5. $-(y + 3)$ _____
6. $-(2x - 9)$ _____

Lesson Quiz
6.5 *Absolute-Value Equations and Inequalities*

Solve each equation. Check your answers.

1. $|x - 1.35| = 0.05$
2. $|x - 8| = 5$
3. $|-4 + x| = 8$

     _____   _____

Solve each inequality. Graph each solution on a number line.

4. $|x + 3| > 8$

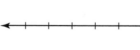

5. $|x + 5| \leq 6$

_____

6. $|4 - x| \geq 5$

_____

Chapter Assessment

Chapter 6, Form A, page 1

Write the letter that best answers the question or completes the statement.

_____ 1. Which inequality corresponds to the statement "b is less than or equal to 1"?

   a. $b \geq 1$ b. $b \leq 1$ c. $b > 1$ d. $b < 1$

_____ 2. Which inequality is true?

   a. $0 \leq -7$ b. $-5 > 1$ c. $1.5 < 1.2$ d. $3.90 > 3.09$

_____ 3. Which sentence represents $n < 4$ on the number line.

   a. Draw a shaded circle at 4. Shade the numbers to the right of 4.
   b. Draw an unshaded circle at 4. Shade the numbers to the left of 4.
   c. Draw an unshaded circle at 4. Shade the numbers to the right of 4.
   d. Draw a shaded circle at 4. Shade the numbers to the left of 4.

_____ 4. Burt wants to buy a pair of shoes that cost \$49.95. He also wants to buy a T-shirt, but he cannot spend more than \$60. Which inequality models this situation?

   a. $49.95 - x > 60$ b. $49.95 + x \geq 60$ c. $x + 49.95 \leq 60$ d. $60 + x \geq 49.95$

_____ 5. Solve $|x - 3.45| = 0.25$.

   a. $x = 3$ and $x = 3.7$ b. $x = -3$ and $x = -3.7$
   c. $x = 3.2$ d. $x = 3.2$ and $x = 3.7$

_____ 6. Solve $\frac{m}{10} - 16 < 40$.

   a. $m < 560$ b. $m > 560$ c. $m < 240$ d. $m > 240$

_____ 7. Solve $-8p < -56$.

   a. $p > 7$ b. $p < 7$ c. $p > -7$ d. $p < -7$

_____ 8. Solve $10 \leq 8 - x$.

   a. $x \leq 2$ b. $x \leq -2$ c. $x \geq 2$ d. $x \geq -2$

_____ 9. Solve $12 + 5x > 7x - 12$.

   a. $x < 12$ b. $x < -12$ c. $x > 12$ d. $x < -12$

_____ 10. Which inequality represents the graph?

   a. $-2 < x < 1$ b. $4 \leq x$ c. $x \leq 3$ or $x > 5$ d. $x > 1$

_____ 11. Solve $-5 \leq 2x - 1 \leq 3$.

   a. $-3 \leq x \leq 2$ b. $x \leq -2$ or $x \geq 2$ c. $x \leq -3$ or $x \geq 2$ d. $-2 \leq x \leq 2$

_____ 12. A company is producing gauges for a contractor. The overhead charges are
       $500. The production costs need to be kept under $1000. How many gauges
       can the company produce at $10 each and keep the costs under $1000?

   a. under 100   b. under 50   c. over 100   d. over 50

_____ 13. Which equation represents the graph?

   a. $y = |x + 2|$ b. $y = |x - 2|$

   c. $y = |x| + 2$ d. $y = |x| - 2$

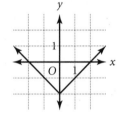

_____ 14. What is the domain and range of the function $y = -|x|$?

   a. Domain: all real numbers; range: all negative numbers and 0
   b. Domain: all negative numbers and 0; range: all real numbers
   c. Domain: all real numbers; range: all positive numbers and 0
   d. Domain: all positive numbers and 0; range: all real numbers

_____ 15. Evaluate $|-20 - 20|$.

   a. -40 b. 40   c. 0   d. not here

_____ 16. Solve $|x + 1| = 5$.

   a. $x = 6$ and $x = -4$ b. $x = 6$ and $x = 4$
   c. $x = -6$ and $x = 4$ d. $x = -6$ and $x = -4$

_____ 17. Solve $|x - 6| \geq 14$.

   a. $-8 \leq x \leq 20$ b. $x \geq 20$ or $x \geq -8$ c. $x \geq 20$ or $x \leq -8$ d. not given

_____ 18. Solve $|x + 8| \geq 1$.

   a. $-9 < x < -7$ b. $-9 > x > -7$ c. $7 < x < 8$ d. no solution

Chapter Assessment

Chapter 6, Form B, page 1

Write an inequality to represent each situation.

1. There were more than 50 people at the party. _____

2. The savings account balance is $450 or less. _____

Solve each inequality.

3. $x + 9 > -1$ _____

4. $y - 6 \leq 5$ _____

5. $3.2 < c + 1.5$ _____

6. $t - \frac{1}{2} > -\frac{1}{4}$ _____

7. In an auditorium that could hold a total of 200 people, 150 people were already seated. Write an inequality to represent the number of people who could still have a seat.

   _____

8. Solve $\frac{a}{7} + 1 > 1.$ _____

9. Solve $-3v + 3 \leq -12.$ _____

10. Is the statement $-8.6 > -8.06$ true or false? _____

11. Write an inequality that corresponds to the following statement:
    t is greater than or equal to negative 25.

    _____

12. Solve $-9z > -36.$ _____

13. Solve $5v + 3 \geq 28.$ _____

14. Solve $\frac{2h}{7} + 1 > 5.$ _____

Chapter Assessment

Chapter 6, Form B, page 2

Evaluate.

15. $|9 - 15|$ _____

16. $|15 - 9|$ _____

17. Solve $|y - 3| = 1.$ _____

18. Mary wants to buy a computer game that costs \$18.50. She also wants to buy a videotape, but she cannot spend more than \$30. Write an inequality to model the situation.

_____

19. Write an inequality that represents the graph.

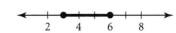

_____

In Exercises 20 and 21, solve each inequality. Show the solution on a number line.

20. $6x > 9$ _____

21. $-\frac{1}{5} - 2 \geq -1$ _____

22. Graph the function $y = |x - 1|$ on the graph at the right.

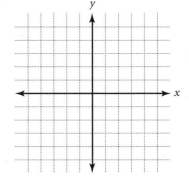

23. Give the domain and range of $y = |x - 1|.$

_____

24. Solve $|x - 2| > 6.$ _____

25. Solve $|x - 7| \leq 2.$ _____

26. Solve $|x - 3.5| = 2.5.$ _____

27. Solve $|-5 + x| = 9.$ _____

28. Solve $2x - 6 \leq 4(x + 3).$ _____

29. Solve $-9 < 2x - 1 < 5.$ _____

Alternative Assessment

Inequalities, Chapter 6, Form A

TASK: To solve problems by using an inequality.

HOW YOU WILL BE SCORED: As you work through the task, your teacher will be looking for the following:

- whether you can write an inequality to describe a real-world situation
- how well you can describe the real-world solutions to an inequality

Chevrolet model	Horsepower (hp)	Wheelbase (in.)	Weight (lb)
Beretta	155	103	2785
Blazer	190	107	4180
Caprice	260	116	4285
Corvette	330	96	3380
Lumina Van	180	110	3890

Source: *Consumer Reports*, April 1996, Volume 61, No. 4, pp. 62–63

1. Find the mean weight of these five models of Chevrolet automobiles.

_____

2. Describe how to use an inequality to find the models whose weight is more than 200 pounds above the mean. Identify them.

_____

_____

3. Write an inequality to represent the horsepower range for the five models of automobiles.

_____

4. Translate the following into an inequality: The wheelbase of the five models is at least 96 inches and at most 116 inches.

_____

5. Describe how you would determine if the following statement is true or false: Each model has at least 155 hp.

_____

6. Discuss how to identify the models that weigh no more than 4000 pounds.

_____

_____

SELF-ASSESSMENT: Describe how you would use profiles of cars to choose the best car for your needs.

Alternative Assessment

Using Compound Absolute-Value Inequalities, Chapter 6, Form B

TASK: To interpret a situation using compound and absolute-value inequalities.

HOW YOU WILL BE SCORED: As you work through the task, your teacher will be looking for the following:

- whether you can write compound inequalities and inequalities using absolute value
- whether you can describe how the solutions to absolute-value inequalities apply to real world situations

The Walker family is planning a weekend trip to the ocean but they have a limited budget. They are looking for a motel with a nightly charge of $98, give or take $10.

1. What is the range of motel prices for one night?

_____

2. Write a compound inequality that shows the range.

_____

3. Write an absolute value inequality describing the range.

_____

4. Describe how you would find the range for the motel cost for the two-night weekend.

_____

5. The Walker's food budget can be described by the inequality $|x - 150| \leq 50$. Describe the Walker's food budget in real world terms.

_____

6. Discuss how to find the Walker family's budget for food and logging in real world terms and write an inequality that describes their budget.

_____

SELF-ASSESSMENT: Compose a compound inequality with an absolute-value inequality.

Quick Warm-Up: Assessing Prior Knowledge

7.1 *Graphing Systems of Equations*

Solve for *y*.

1. $4x - y = 7$ _____

2. $x + 2y = 8$ _____

Lesson Quiz

7.1 *Graphing Systems of Equations*

Solve by graphing. Check by substituting the solutions into the original equations.

1. $\begin{cases} 3x + 2y = 8 \\ 6x - 4y = 8 \end{cases}$

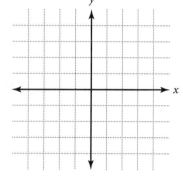

2. $\begin{cases} 2x + y = -4 \\ 2y = 2x + 10 \end{cases}$

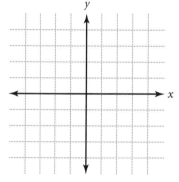

_____   _____

3. Algebraically determine whether the point $(2, -1)$ is a solution to the system of equations below. Then check by graphing the system on the grid provided.

$\begin{cases} y = x - 3 \\ y = 2x - 5 \end{cases}$

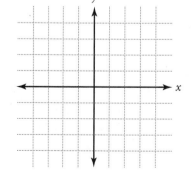

_____

Quick Warm-Up: Assessing Prior Knowledge
7.2 The Substitution Method

Simplify.

1. $4r + 3(r + 5)$ _____
2. $z - 2(z - 2)$ _____
3. $5(4p - 7) - 2p$ _____

Solve for y.

4. $2x + y = 5$ _____
5. $6x - 2y = 12$ _____

Lesson Quiz
7.2 The Substitution Method

Solve by using substitution and check your answers.

1. $\begin{cases} 5x + 3y = 4 \\ y = -3x \end{cases}$ _____

2. $\begin{cases} x = 2y \\ x - y = 1 \end{cases}$ _____

3. $\begin{cases} x + 2y = 6 \\ x - 3y = 1 \end{cases}$ _____

4. $\begin{cases} x + y = -7 \\ 2x - 8y = 6 \end{cases}$ _____

5. $\begin{cases} 2x - y = 5.4 \\ x + y = 1.2 \end{cases}$ _____

6. $\begin{cases} 2x + y = 2 \\ y = 3x - 4 \end{cases}$ _____

Write and solve a system of equations for each problem.

7. The sum of two numbers is 52. The smaller number is 12 less than the larger number. Find the two numbers. _____

8. The perimeter of a rectangle is 120 meters. Its length is 3 times its width. Find the dimensions of the rectangle. _____

Quick Warm-Up: Assessing Prior Knowledge
7.3 The Elimination Method

Simplify.

1. $(6x - y) + (-7x + y)$ _____

2. $-4(x + 3y)$ _____

Solve each system.

3. $\begin{cases} 5x - 2y = 3 \\ y = 2x \end{cases}$ _____

4. $\begin{cases} 2x + y = -1 \\ y = -3x - 3 \end{cases}$ _____

Lesson Quiz
7.3 The Elimination Method

Choose a method for solving each system, and explain your decision. Do not solve.

1. $\begin{cases} 5x + 3y = -4 \\ 2x - 6y = 20 \end{cases}$ _____

2. $\begin{cases} 4x + 7y = 19 \\ y = x + 9 \end{cases}$ _____

3. $\begin{cases} y = 3x - 12 \\ y = 8x - 12 \end{cases}$ _____

Solve each system of equations by elimination, and check your solution.

4. $\begin{cases} 4x + 5y = 7 \\ 2x + 3y = 5 \end{cases}$ _____

5. $\begin{cases} 2x - 4y = 14 \\ x = 6y + 11 \end{cases}$ _____

Solve each system of equations by using any method.

6. $\begin{cases} 3x + \frac{1}{3}y = 10 \\ 2x - 5 = \frac{1}{3}y \end{cases}$ _____

7. $\begin{cases} x + 2y = 20 \\ 3x - 4y = -10 \end{cases}$ _____

8. $\begin{cases} 5x - 3y = 7 \\ 3x - y = 1 \end{cases}$ _____

9. $\begin{cases} y = 3x - 1 \\ 2x + y = 9 \end{cases}$ _____

Mid-Chapter Assessment

Chapter 7 (Lessons 7.1–7.3)

Write the letter that best answers the question or completes the statement.

_____ 1. Which system is graphed at right?

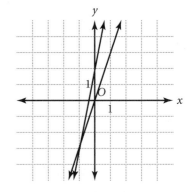

    a. $\begin{cases} x - 3y = -1 \\ y = x - 2 \end{cases}$ b. $\begin{cases} x + 2y = -1 \\ 3x - 2y = -3 \end{cases}$

    c. $\begin{cases} 3x - y = 0 \\ y - 5x = 2 \end{cases}$ d. $\begin{cases} \frac{1}{3}x + y = -4 \\ x - y = 0 \end{cases}$

_____ 2. What is the point of intersection of $x + y = 3$
    and $x + 5y = 7$?

    a. $(-2, 1)$ b. $(2, 1)$ c. $(-3, 2)$ d. $(2, -3)$

_____ 3. Choose the best method for solving the system $\begin{cases} 120x - y = 100 \\ x + y = 21 \end{cases}$.

    a. graphing   b. substitution   c. elimination   d. technology

_____ 4. Which represents the solution to the system $\begin{cases} 3x + \frac{2}{3}y = 4 \\ 4x + \frac{2}{3}y = 6 \end{cases}$

    a. $(-3, 2)$ b. $(3, 4)$ c. $(4, 6)$ d. $(2, -3)$

Solve each system of equations by using any method.

5. $\begin{cases} y = x + 3 \\ 3y + x = 5 \end{cases}$ _____

6. $\begin{cases} y = 3x \\ x + 2y = -21 \end{cases}$ _____

7. $\begin{cases} 2x - 3y = -4 \\ x = 7 - 3y \end{cases}$ _____

8. $\begin{cases} 0.5x + 0.2y = 3 \\ 2x - 3y = -7 \end{cases}$ _____

9. Carol bought 5 blouses and 3 skirts for a total of $215. At the
same time, her sister Eileen bought 3 blouses and 2 skirts for
a total of $135. What was the price of each blouse and each skirt?

_____

Quick Warm-Up: Assessing Prior Knowledge
7.4 Consistent and Inconsistent Systems

Solve.

1. $5a + (7 - 2a) = 4a - 1$ _____

2. $9z - 2(2z - 3) = 6z + 6 - z$ _____

3. $-4n - 2 + 3n = 8 - n + 6$ _____

Lesson Quiz
7.4 Consistent and Inconsistent Systems

1. Use the equation $y = 3x - 5$ and another equation
   to create a system that is dependent.   _____

   _____

2. Use the equation $y = \frac{1}{2}x + 3$ and another equation
   to create a system that is inconsistent.   _____

   _____

Tell if the given system is consistent or inconsistent.

3. $\begin{cases} 2a - 3b = -1 \\ 5a - 7.5b = -2.5 \end{cases}$ _____

4. $\begin{cases} y = -3x - 8 \\ y = -3x + 8 \end{cases}$ _____

Solve each system algebraically.

5. $\begin{cases} 3x + 8y = 14 \\ \frac{3}{2}x + 4y = 7 \end{cases}$ _____

6. $\begin{cases} 3x + 7y = 5 \\ 3y = -2x \end{cases}$ _____

7. $\begin{cases} x - 5y = 13 \\ 2x + y = 15 \end{cases}$ _____

8. $\begin{cases} 4x + 5y = 10 \\ y = 0.8 - 1 \end{cases}$ _____

Quick Warm-Up: Assessing Prior Knowledge
7.5 Systems of Inequalities

Solve each inequality. Graph the solution on a number line.

1. $-3s \geq 3$ _____

2. $-m + 2 < 4m - 8$ _____

3. $b + 4 < 2$ or $6 < 2b$ _____

4. $-9 \leq 2q - 1 < 5$ _____

Lesson Quiz
7.5 Systems of Inequalities

1. Describe the boundary line for the graph of $x + 3y \geq 6$.

_____

2. What region would you shade for the graph of $y - x \leq 3$?

_____

Solve by graphing.

3. $\begin{cases} y < x + 2 \\ y > -x - 1 \end{cases}$

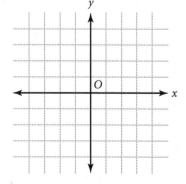

4. $\begin{cases} y > x - 2 \\ x \leq 3 \end{cases}$

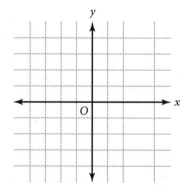

5. The perimeter of triangle ABC is at most 10 centimeters. The perimeter of rectangle $PQRS$ is at least 4 centimeters. Write a system of inequalities defined by the perimeters.

_____

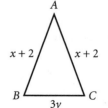

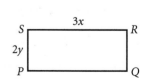

Quick Warm-Up: Assessing Prior Knowledge
7.6 Classic Puzzles in Two Variables

Write an equation to model each situation.

1. A person drove 150 miles in 3 hours at an average rate of *r* miles

   per hour. _____

2. An 8-ounce mixture of nuts consists of *p* ounces of peanuts and

   c ounces of cashews. _____

3. A collection of *q* quarters is worth a total of $11.75. _____

Lesson Quiz
7.6 Classic Puzzles in Two Variables

1. Peggy has 500 milliliters of a 60% solution of copper sulfate. How
   many milliliters of a 30% solution of copper sulfate should be
   added to obtain a 50% solution?   _____

2. Bill has 6 more quarters than nickels. The coins have a total value of
   $6. How many of each kind of coin does Bill have?   _____

3. The sum of the digits of a two-digit number is 11. The original
   two-digit number is 5 more than 3 times the number with
   its digits reversed. Find the original number.   _____

4. A boat can travel 10 miles downstream in 2 hours and can travel
   the same distance upstream in $3\frac{1}{3}$ hours. Find the rate of the
   boat in still water and the rate of the current.

   _____

5. Judy is 3 times as old as Phyllis. In 10 years, Judy will be twice
   as old as Phyllis. Find their present ages.

   _____

Chapter Assessment

Chapter 7, Form A, page 1

Write the letter that best answers the question or completes the statement.

_____ 1. Which system is graphed at right?

a. $\begin{cases} x + 2y = 3 \\ y = -2x - 1 \end{cases}$ b. $\begin{cases} x + 2y = 3 \\ x + 2y = -1 \end{cases}$

c. $\begin{cases} 2x + y = 3 \\ 6x + 3y = 9 \end{cases}$ d. $\begin{cases} 2x + y = 3 \\ y = -2x - 1 \end{cases}$

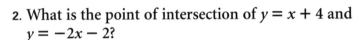

_____ 2. What is the point of intersection of $y = x + 4$ and $y = -2x - 2$?

a. $(-2, 2)$ b. $(1, -4)$ c. $(1, 5)$ d. $(2, -2)$

_____ 3. Which of the following is the best method for solving the system?

$$\begin{cases} x + \frac{2}{3}y = 2 \\ \frac{1}{2}x - \frac{1}{3}y = -3 \end{cases}$$

a. graphing b. substitution c. elimination d. technology

_____ 4. Which system is graphed at right?

a. $\begin{cases} 2x + 4y \leq 4 \\ 4x + 2y \geq 2 \end{cases}$ b. $\begin{cases} y \leq 2x - 4 \\ y \geq \frac{1}{2}x - 2 \end{cases}$

c. $\begin{cases} y \geq 2x - 4 \\ y \leq \frac{1}{2}x - 2 \end{cases}$ d. $\begin{cases} 2x + 4y \geq -4 \\ 4x + 2y \leq -2 \end{cases}$

_____ 5. Solve the system $\begin{cases} y = 2x \\ x + y = 12 \end{cases}$.

a. $(4, 8)$ b. $(2, 12)$ c. $(8, 4)$ d. $(1, 2)$

_____ 6. Solve the system $\begin{cases} y = \frac{1}{3}x + 6 \\ 3y - x = 24 \end{cases}$.

a. $\left(\frac{1}{3}, 3\right)$ b. $(6, 0)$ c. $(0, 2)$ d. no solution

_____ 7. Solve the system $\begin{cases} 3x + 2y = 13 \\ 5x = 2y + 11 \end{cases}$.

a. $(3, 2)$ b. $(5, 2)$ c. $(1, 5)$ d. no solution

Chapter Assessment

Chapter 7, Form A, page 2

_____ 8. Solve the system $\begin{cases} 6x - 2y = 6 \\ x + 2y = 8 \end{cases}$.

    a. $(1, 0)$ b. $(-1, -6)$ c. $(2, 3)$ d. infinite solutions

_____ 9. Find the value of k in the inconsistent system $\begin{cases} 2x + y = 8 \\ y = kx - 3 \end{cases}$.

    a. -2 b. 2 c. $-\dfrac{1}{2}$ d. $\dfrac{1}{2}$

_____ 10. Find the equation of a line that would form an independent system with $y = 3x - 5$ and that contains the point $(2, 1)$.

    a. $y = 3x + 5$ b. $2y - 6x = 10$ c. $2x = y + 3$ d. $y = 2x + 1$

_____ 11. Cathy has $2.30 in dimes and quarters. If the number of dimes is 5 less than the number of quarters, how many coins of each type does she have?

    a. 8 dimes, 6 quarters   b. 3 dimes, 8 quarters

    c. 8 dimes, 3 quarters   d. 3 dimes, 5 quarters

_____ 12. The sum of the digits of a two-digit number is 13. If the digits are reversed, the new number is 4 less than twice the original number. Find the original number.

    a. 58   b. 67   c. 94   d. 49

_____ 13. Solve the system $\begin{cases} x + 3y = 15 \\ y = -\dfrac{1}{3}x + 1 \end{cases}$.

    a. $(0, 5)$ b. $(3, 4)$ c. $(-3, 6)$ d. no solution

_____ 14. A solution of 60 milliliters is 50% acid. How many milliliters of water must be added to create a 30% acid solution?

    a. 40 mL   b. 80 mL   c. 48 mL   d. 24 mL

_____ 15. A glider can fly 240 kilometers in 3 hours with the wind. But against the wind, the glider can only fly 210 km in 3 hours. Find the speed of the glider and the speed of the wind.

    a. glider: 80 km/h   b. glider: 75 km/h
        wind: 10 km/h   wind: 5 km/h

    c. glider: 45 km/h   d. glider: 75 km/h
        wind: 45 km/h   wind: 10 km/h

Chapter Assessment
Chapter 7, Form B, page 1

1. Explain how to write an equation that would form a dependent system with the equation $2x - y = 5$.

   _____

2. Describe the boundary line for the graph of $x + 3y > 4$.

   _____

Graph each system on the grid provided to find a solution.
Check your answer algebraically.

3. $\begin{cases} 2x + 4y = -4 \\ y = x + 2 \end{cases}$

4. $\begin{cases} 3x - y = 3 \\ \frac{1}{3}y = x - 1 \end{cases}$

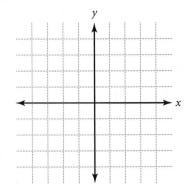

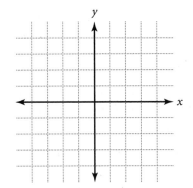

_____   _____

Solve by using the best method for each system.
Check your solutions.

5. $\begin{cases} 2x + 3y = 7 \\ x + 2y = 4 \end{cases}$ _____

6. $\begin{cases} 4x - y = 11 \\ 2x + y = 1 \end{cases}$ _____

7. $\begin{cases} 3x + 2y = 1 \\ 6x + 5y = 4 \end{cases}$ _____

8. $\begin{cases} 3x - 3y = 18 \\ x + y = 3 \end{cases}$ _____

9. $\begin{cases} x - 4y = 5 \\ 2x + 4y = -2 \end{cases}$ _____

10. $\begin{cases} 6x - 2y = 6 \\ y = 3x + 2 \end{cases}$ _____

11. The sum of two numbers is 48. If the larger number is 6 less than twice the smaller number, find the two numbers. _____

12. The perimeter of a rectangle is 64 meters. If the length is 2 more than twice the width, find the dimensions of the rectangle. _____

Chapter Assessment

Chapter 7, Form B, page 2

Identify each system as *consistent and independent, consistent
and dependent,* or *inconsistent.*

13. $\begin{cases} 3x + 2y = 6 \\ 9x + 6y = 18 \end{cases}$ _____

14. $\begin{cases} y + 2x = 10 \\ y - 2x = 6 \end{cases}$ _____

15. $\begin{cases} y - 3x = 1 \\ y = 3x - 1 \end{cases}$ _____

Answer the following questions by referring to the graph:

16. Write the equation of each line:

$\overleftrightarrow{AB}$ _____

$\overleftrightarrow{CD}$ _____

17. Write the equation of a line that would form an
inconsistent system with line *AB*.

_____

18. Write the equation of a line that would form a
dependent system with line *CD*.

_____

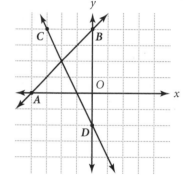

Graph the solution set of each system on the grid provided.

19. $\begin{cases} y \le x + 3 \\ y \ge x + 2 \end{cases}$

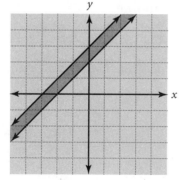

20. $\begin{cases} y - 2x > 2 \\ 2y - x < -2 \end{cases}$

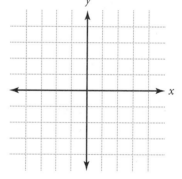

21. A parking meter contains $8.25 in dimes and quarters. If the
number of quarters is 9 more than twice the number of dimes, how
many of each coin is in the parking meter? _____

22. The sum of the ages of Jon and his brother is 40. Jon's age is 8 more
than his brother's age. Find the age of each. _____

23. Traveling with the wind, a plane takes $2\frac{1}{2}$ hours to fly a distance
of 1500 miles. The return flight against the wind takes 3 hours.
Find the speed of the plane and the speed of the wind. _____

Alternative Assessment

Exploring Systems of Equations, Chapter 7, Form A

TASK: Describe how to solve systems of equations by graphing.

HOW YOU WILL BE SCORED: As you work through the task, your teacher will be looking for the following:

- whether you can solve systems of equations by graphing
- how well you can describe the process of graphing a solution
- whether you can solve systems of equations by using elimination

Solve each system by graphing.

1. $\begin{cases} x - y = 4 \\ y = -3 \end{cases}$

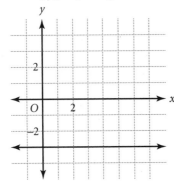

2. $\begin{cases} y = 2x + 1 \\ y = -3x - 4 \end{cases}$

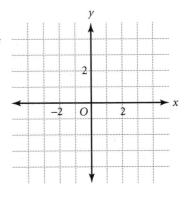

_____   _____

3. Describe how to solve a system of equations by graphing.

_____

_____

Solve each system of equations by elimination.

4. $\begin{cases} 3x + y = 4 \\ x - y = 4 \end{cases}$

5. $\begin{cases} 3x - 8y = 2 \\ -x + 2y = 2 \end{cases}$

6. $\begin{cases} 2x - y = 3 \\ 2x - y = -2 \end{cases}$

_____   _____   _____

7. How can you use the substitution method to check the solution for the system of equations in Exercise 4?

_____

SELF-ASSESSMENT: What does it mean when a system of equations has no solution?

Alternative Assessment

Formulating Problems, Chapter 7, Form B

TASK: Formulate your own coin, digit, mixture, and motion problems.

HOW YOU WILL BE SCORED: As you work through the task, your teacher will be looking for the following:

- how well you can work backward to formulate your own problems
- whether you can write and solve a system of equations for your problems

1. Write a digit problem that has an answer of 48. Write a system of equations for your problem. Then find the solution.

   _____

   _____

   _____

2. Write a coin problem that has an answer of 21 quarters and 24 dimes. Write a system of equations for your problem. Then find the solution.

   _____

   _____

   _____

3. Write a mixture problem that has answers of 240 grams of a 15% solution and 160 grams of a 40% solution. Write a system of equations for your problem. Then find the solution.

   _____

   _____

   _____

4. Write a motion problem that has an answer of 24 miles per hour. Write a system of equations for your problem. Then find the solution.

   _____

   _____

   _____

SELF-ASSESSMENT: Explain how you formulated one of your problems. What was difficult about creating your own problem? How did you work through that difficulty?

Quick Warm-Up: Assessing Prior Knowledge
8.1 *Laws of Exponents: Multiplying Monomials*

Evaluate.

1. $2 \cdot 2 \cdot 2 \cdot 2$ _____

2. $(2)(3)(4)$ _____

3. $(2)(-3)(4)$ _____

4. $(-2)(-3)(-4)$ _____

5. $(-2)(3)(-4)$ _____

6. $(-14)(-7)$ _____

Lesson Quiz
8.1 *Laws of Exponents: Multiplying Monomials*

Find the value of each expression.

1. 5^3 _____

2. 2^4 _____

3. 7^2 _____

4. 10^2 _____

5. 2^5 _____

6. 6^2 _____

7. 9^3 _____

8. 3^5 _____

9. 12^1 _____

10. 4^4 _____

11. 10^4 _____

12. 1^8 _____

Simplify each product. Leave the product in exponent form.

13. $10^3 \cdot 10^5$ _____

14. $2^3 \cdot 2^3$ _____

15. $5^2 \cdot 5^3$ _____

16. $4^5 \cdot 4^3$ _____

17. $10^2 \cdot 10^5$ _____

18. $7^5 \cdot 7^4$ _____

19. $(4a^2)(3a^3)$ _____

20. $(2b^4)(-b^2)$ _____

21. $(6mn^2)(3m^3)$ _____

22. $(-5r^3)(-2r^2s)$ _____

23. $(x^2y^3)(x^3y^4)$ _____

24. $(8a^2b^2c^3)(4ab^4c^2)$ _____

25. The area, A, of a rectangle can be found by using the formula $A = lw$, where l is the length and w is the width of the rectangle. Find the area of a rectangle that has a length of $2x$ and a width of $3xy^2$. _____

Quick Warm-Up: Assessing Prior Knowledge

8.2 *Laws of Exponents: Powers and Products*

Evaluate each expression.

1. a^2 for $a = 7$ _____

2. $\frac{1}{b^2}$ for $b = 3$ _____

3. c^2d for $c = 4$ and $d = 5$ _____

4. $(fg)^2$ for $f = 4$ and $g = 2$ _____

5. $h(j + k)^2$ for $h = 2, j = 3,$ and $k = 1$ _____

Lesson Quiz

8.2 *Laws of Exponents: Powers and Products*

Simplify each of the following.

1. $(4^2)^3$ _____

2. $(10^3)^2$ _____

3. $(2^4)^3$ _____

4. $(x^3)^3$ _____

5. $(-y^2)^3$ _____

6. $(-5t)^3$ _____

7. $(b^2)^3$ _____

8. $2(s)^2$ _____

9. $(5m^2)^2$ _____

10. $(3ab^3)^2$ _____

11. $(-2r^2s)^4$ _____

12. $(-3pt)^3$ _____

13. $(4c^2d)^2$ _____

14. $4(m^2n)^3$ _____

15. $-(y^2z^3)^2$ _____

16. $(-5ct^2)^3$ _____

17. $(8e^4f^5)^2$ _____

18. $-5(-g^3h^5)^3$ _____

19. $(a^2)(a^3)^2$ _____

20. $(ab^2)^3(b^2)^2$ _____

21. $(cd)^3(c^2d)^4$ _____

22. $(-p^3q^2)^4(p^2)^2$ _____

23. $5(b^4c)^2(c^2)^3$ _____

24. $(-2r^5s)^2(-3rs^2)^3$ _____

25. The area, *A*, of a square can be found by using the formula
$A = s^2$, where *s* is the length of one of the sides of the
square. Find the area of a square with side lengths of c^2d^3. _____

NAME _____ CLASS _____ DATE _____

Quick Warm-Up: Assessing Prior Knowledge
8.3 *Laws of Exponents: Dividing Monomials*

Evaluate.

1. $54 \div (-6)$ _____

2. $-24 \div (-4)$ _____

3. 2^5 _____

4. 5^3 _____

Simplify.

5. $(m^3)^6$ _____

6. $(y^r)^4$ _____

7. $(3a^2)^5$ _____

8. $(p^2q^3)^2$ _____

Lesson Quiz
8.3 *Laws of Exponents: Dividing Monomials*

Use the Quotient-of-Powers Property to simplify each quotient. Then find the value of the result.

1. $\dfrac{5^7}{5^3}$ _____

2. $\dfrac{3^4}{3^3}$ _____

3. $\dfrac{2^{10}}{2^7}$ _____

Use the Quotient-of-Powers Property to simplify each quotient.

4. $\dfrac{m^7}{m^2}$ _____

5. $\dfrac{9c^4}{c}$ _____

6. $\dfrac{a^p}{a^q}$ _____

Simplify each expression.

7. $\left(\dfrac{r}{s}\right)^3$ _____

8. $\left(\dfrac{5c}{d^2}\right)^2$ _____

9. $\left(\dfrac{9g^7}{3g^2}\right)^2$ _____

10. $\left(\dfrac{n^2p}{n}\right)^3$ _____

11. $\left(\dfrac{-5d^2z^6}{10dz^3}\right)^2$ _____

12. $\left(\dfrac{j^2k^5}{3}\right)^w$ _____

Find each quotient. Assume that the conditions of the Quotient-of-Powers Property are met.

13. $\dfrac{15c^6}{3c}$ _____

14. $\dfrac{-6t^8s^6}{18t^5s^2}$ _____

15. $\dfrac{-21u^7v^8}{-28u^6v}$ _____

16. $\dfrac{3.2(a^4b)^2}{8a^3}$ _____

17. $\dfrac{2.5d^8e^6}{(2.5d^2)^3}$ _____

18. $\dfrac{-36b^2(bd^2)^3}{(6b)^2d^3}$ _____

Quick Warm-Up: Assessing Prior Knowledge
8.4 *Negative and Zero Exponents*

Evaluate.

1. $3^2 \cdot 3^4$ _____

2. $2^6 \cdot 2$ _____

3. $\dfrac{9^{13}}{9^{11}}$ _____

4. $\dfrac{2^{12}}{2^6}$ _____

Simplify.

5. $k^4 \cdot k^8$ _____

6. $(-2t)^3$ _____

7. $\dfrac{a^4b^7}{ab^6}$ _____

8. $\dfrac{8v^8}{2v^2}$ _____

Lesson Quiz
8.4 *Negative and Zero Exponents*

Match each expression with the letter that indicates the equivalent value.

1. $(-2)^3$ _____

2. 2^{-3} _____

3. 2^3 _____

4. -2^{-3} _____

a. -8 b. $-\dfrac{1}{8}$ c. $\dfrac{1}{8}$ d. 8

Simplify and write each expression with positive exponents only.

5. -3^{-2} _____

6. x^2y^{-3} _____

7. $a^{-5}b^4$ _____

8. $c^{-1}d^0$ _____

9. $(p^3q^{-5})^0$ _____

10. $3t^{-5}t^2$ _____

11. $\dfrac{15m^4}{3m^{-2}}$ _____

12. $\dfrac{r^{-2}}{2r^{-3}}$ _____

13. $\dfrac{(2x^2)^3(10x^{-4})}{8x^{-1}}$ _____

14. $-a^{-3}a^{-1}$ _____

15. 10^{-7} _____

16. $\dfrac{m^2}{m^{-5}}$ _____

Evaluate each of the following mentally:

17. $(23.71 \times 0)^{19}$ _____

18. $(0.637 \times 9.005)^0$ _____

19. $\dfrac{37.89^{12}}{37.89^{13}}$ _____

Mid-Chapter Assessment
Chapter 8 (Lessons 8.1–8.4)

Write the letter that best answers the question or completes the statement.

_____ 1. What is the value of $(10^3 \cdot 10^{-4})^2$?

  a. 1000 **b.** $\frac{1}{10}$ **c.** $\frac{1}{100}$ **d.** 100

_____ 2. What is the value of $\frac{3^2}{3^{-1}}$?

  a. 27 **b.** 24 **c.** 9 **d.** 3

_____ 3. The product of $(-5a^3b)(-ab^2)^3$ is

  a. $-5a^6b^6$. **b.** $5a^7b^6$. **c.** $-5a^4b^3$. **d.** $5a^6b^7$.

_____ 4. The quotient $\frac{(3x^2y^{-1})^3}{18x^3y^2}$ is equivalent to

  a. $\frac{3x^3}{2y^5}$. **b.** $\frac{3}{2x^2y}$. **c.** $\frac{1}{6xy^5}$. **d.** $\frac{x^2}{6y}$.

_____ 5. What is the value of $(10a)^3$ when a is 2?

  a. 1000 **b.** 8000 **c.** 100 **d.** 800

Simplify each of the following:

6. $\frac{5^3 + 5^2}{2(5)^2}$ _____

7. $\frac{10^2(10^0)^3}{10^{-3}}$ _____

8. $\frac{10(-4a^5b)^2}{2a^0b^2}$ _____

9. $\frac{5x^3y^4}{(5x^2)^2y^{-1}}$ _____

10. Explain why $(x^a)^b = (x^b)^a$ is true. _____

_____

Quick Warm-Up: Assessing Prior Knowledge
8.5 *Scientific Notation*

Evaluate.

1. $5^5 \cdot 5^{-1}$ _____

2. $10^4 \cdot 10^{-6}$ _____

3. $\dfrac{10^7}{10^3}$ _____

4. $\dfrac{3^5}{3^7}$ _____

5. $(3.2)(100)$ _____

6. $(3.2)(0.01)$ _____

Lesson Quiz
8.5 *Scientific Notation*

1. Explain what happens to a number when you move its decimal point
   8 places to the right.

   _____

Write each number in scientific notation.

2. 3,000,000 _____

3. 720,000 _____

4. 63,400,000 _____

5. 0.008 _____

6. 0.000075 _____

7. 0.000000905 _____

Write each number in decimal notation.

8. 5×10^5 _____

9. 7×10^{-8} _____

10. 8.3×10^{-1} _____

11. 6.02×10^7 _____

12. 2.907×10^{10} _____

13. 1.5×10^{-6} _____

**Perform the following computations. Write your answers in
scientific notation.**

14. $(4 \times 10^5)(2 \times 10^3)$ _____

15. $(7 \times 10^4)(3 \times 10^7)$ _____

16. $(9 \times 10^8)(5 \times 10^{-15})$ _____

17. $\dfrac{(5 \times 10^6)(8 \times 10^3)}{2 \times 10^7}$ _____

18. $\dfrac{18 \times 10^7}{4 \times 10^{12}}$ _____

19. $\dfrac{(2 \times 10^3)(5 \times 10^8)}{2 \times 10^5}$ _____

Quick Warm-Up: Assessing Prior Knowledge

8.6 *Exponential Functions*

Evaluate.

1. 3^0 _____

2. 3^1 _____

3. 3^2 _____

4. 3^3 _____

5. 3^4 _____

6. 3^5 _____

7. $(0.5)^0$ _____

8. $(0.5)^1$ _____

9. $(0.5)^2$ _____

10. $(0.5)^3$ _____

11. $(0.5)^4$ _____

12. $(0.5)^5$ _____

Lesson Quiz

8.6 *Exponential Functions*

1. What is the general growth formula? _____

Graph each of the following on the grids provided:

2. $y = 3^x$

3. $y = 3^x + 2$

4. $y = 3^{x+2}$

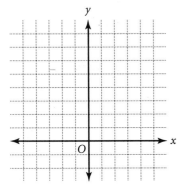

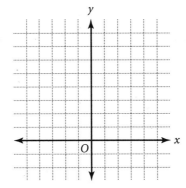

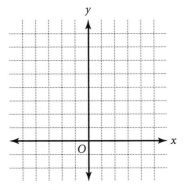

Let $y = 3^x$. Find the values of y for the following values of x:

5. $x = 2$ _____

6. $x = 0$ _____

7. $x = -1$ _____

8. $x = -2$ _____

The population of France was about 57.3 million in 1992 and was growing at a rate of about 0.4% per year. Use this information to estimate the following:

9. the population of France in 1995 _____

10. the population of France in 2001 _____

Quick Warm-Up: Assessing Prior Knowledge
8.7 *Applications of Exponential Functions*

Evaluate.

1. x^3 for $x = 4$ _____

2. 3^x for $x = 4$ _____

3. $3y^z$ for $y = 2$ and $z = 4$ _____

4. $7(1 + r)^2$ for $r = 5$ _____

5. $4(1 + r)^t$ for $r = 2$ and $t = 6$ _____

Lesson Quiz
8.7 *Applications of Exponential Functions*

1. What is the formula for exponential growth? _____

2. What is the formula for exponential decay? _____

An investment of $1000 is growing at 5% per year. Find the value of the investment at the end of:

3. year 1. _____

4. year 2. _____

5. year 3. _____

6. year 4. _____

7. year 5. _____

8. year x. _____

The population of Greece was 10,064,000 in 1992 and was increasing at the rate of 0.2% a year.

9. Estimate the population 10 years from 1992. _____

10. Estimate the population 5 years prior to 1992. _____

Chapter Assessment

Chapter 8, Form A, page 1

Write the letter that best answers the question or completes the statement.

 1. What is the value of $\frac{c^4 - c^2}{c^{-1}}$ when $c = 3$?

    a. 3   b. 24   c. 27   d. 216

_____ 2. In customary notation "three billion, five hundred thirty million" is

    a. 3,530,000,000.   b. 3,500,030,000.   c. 3,530,000.   d. 350,030.

_____ 3. In simplified form, $(5x^{-2}y^3)(-xy^2)$ is

    a. $-5x^3y^5$.   b. $-\frac{5y^5}{x}$.   c. $-\frac{y^5}{5x}$.   d. $-5xy^5$.

_____ 4. In simplified form, $\frac{-3a^2b^6}{-18a^5b^2}$ is

    a. $-\frac{b^3}{6a^3}$.   b. $6a^3b^4$.   c. $-\frac{b^4}{6a^3}$.   d. $\frac{b^4}{6a^3}$.

_____ 5. In simplified form, $(3c^2d^5)(2c^3d^0)^2$ is

    a. $6c^8d^7$.   b. $12c^7d^5$.   c. $12c^8d^5$.   d. $6c^{12}d$.

_____ 6. In simplified form, $(v^{-3}w^3)(v^{-3}w^2)$ is

    a. w.   b. $\frac{w^5}{v^6}$.   c. v^6w^5.   d. $\frac{w}{v}$.

_____ 7. In simplified form, $(-3)^{-3}$ is

    a. $-\frac{1}{27}$.   b. -27.   c. 27.   d. $\frac{1}{27}$.

_____ 8. In scientific notation, 28,700,000,000 is written as

    a. 2.87×10^7.   b. 287×10^7.   c. 2.87×10^{10}.   d. 2.87×10^9.

_____ 9. In decimal notation, 6.05×10^{-9} is written as

    a. 0.000000000605   b. 0.00000000605   c. 0.00000605   d. 0.000000605

Chapter Assessment

Chapter 8, Form A, page 2

In Exercises 10–12, find the value of each expression.

_____ 10. $(2.5 \times 10^6)(6 \times 10^5)$

      a. 150×10^{11} b. 1.5×10^{31} c. 15×10^{30} d. 1.5×10^{12}

_____ 11. $\dfrac{7 \times 10^3}{2 \times 10^7}$

      a. 3.5×10^4 b. 3.5×10^{-4} c. 3.5×10^{10} d. 3.5×10^{-5}

_____ 12. $\dfrac{(3 \times 10^6)(8 \times 10^4)}{6 \times 10^8}$

      a. 4×10^2 b. 4×10^3 c. 5×10^2 d. 2.4×10^5

_____ 13. Which function is graphed below?

      a. $y = \left(\dfrac{1}{4}\right)^x$

      b. $y = 4^x$

      c. $y = 3^x$

      d. $y = \left(\dfrac{1}{3}\right)^x$

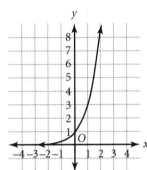

_____ 14. The population of a country was about 57.9 million in 1998 and was growing at a rate of about 0.1% per year. What is the estimated population of this country for 2005?

      a. 58.2 million   b. 58.3 million   c. 62 million   d. 62.3 million

_____ 15. An investment has been growing at a rate of 7% per year and now has a value of $9500. What was the value of the investment, to the nearest dollar, 5 years ago?

      a. $669   b. $6751   c. $6773   d. $13,324

Chapter Assessment

Chapter 8, Form B, page 1

Find the value of each expression.

1. $(2^3)^2$ _____

2. $\dfrac{10^3 \cdot 10^8}{10^5}$ _____

3. $\dfrac{3^3 + 3^3}{6^2}$ _____

Simplify each of the following:

4. $(5p^4)(6p^7)$ _____

5. $(-3a^3b^2)(-7a^5b)$ _____

6. $(-4c^6d)(2cd^2)^{-3}$ _____

7. $(5x^{-1}y^3)^2(3x^4y^{-5})$ _____

8. $\dfrac{42r^6}{7r^9}$ _____

9. $\dfrac{-15v^5w^3}{25v^7w^0}$ _____

10. $\dfrac{3.5ab^{-2}}{7a^{-1}b^3}$ _____

11. $\dfrac{-(-2x^2y^{-1})^2}{4x^3y^2}$ _____

12. Explain why $(5c)^{-3}$ does not equal $\dfrac{125}{c^3}$.

_____

Each of the following numbers is given in calculator notation. Write the numbers in decimal notation.

13. 3.08 E 05 _____

14. -2 E -04 _____

Write each number in scientific notation.

15. 90,000,000 _____

16. 305 _____

17. 0.00007 _____

18. 0.00046 _____

19. 12,800,000 _____

20. 0.00000609 _____

Write each number in decimal notation.

21. 8×10^6 _____

22. 7.2×10^5 _____

23. 1.05×10^7 _____

24. 3×10^{-5} _____

25. 2.06×10^{-9} _____

26. 5.1×10^{-4} _____

Chapter Assessment

Chapter 8, Form B, page 2

Perform the following computations. Write your answers in scientific notation.

27. $(3 \times 10^5)(2 \times 10^3)$ _____

28. $(7 \times 10^8)(5 \times 10^6)$ _____

29. $\dfrac{7 \times 10^9}{2 \times 10^4}$ _____

30. $\dfrac{4 \times 10^{12}}{8 \times 10^5}$ _____

31. $\dfrac{(6 \times 10^7)(3 \times 10^5)}{9 \times 10^6}$ _____

32. $(4 \times 10^{11})(5 \times 10)(2 \times 10^{-10})$ _____

Graph each of the following:

33. $y = 0.5^x - 1$

34. $y = 2^x$

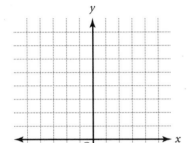

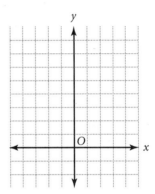

An investment has been losing money at a rate of 3% per year and now has a value of $75,000. Find the value of the investment to the nearest dollar.

35. 5 years from now _____

36. 10 years ago _____

An investment is growing at a rate of 7% per year and now has a value of $12,000. Find the value of the investment (to the nearest dollar).

37. in five years _____

38. ten years ago _____

The population of a country was estimated to be 119,000,000 in 1999 and was growing at the rate of 2.3% a year. Estimate, to the nearest million, the population in

39. the year 2003. _____

40. the year 2007. _____

Alternative Assessment

Laws of Exponents, Chapter 8, Form A

TASK: Use the properties of exponents to evaluate powers of monomials.

HOW YOU WILL BE SCORED: As you work through the task, your teacher will be looking for the following:

- whether you can use the properties of exponents to simplify monomials
- how well you understand the process of computing powers of monomials

1. Write a statement about the property of exponents needed to simplify $\frac{2ab^3}{b}$. Then find the quotient.

_____

_____

2. Describe how to use the Product-of-Powers Property to simplify $(5x^2y)(-xy)$. Then find the product.

_____

_____

3. Explain how the Power-of-a-Product-Property can be used to evaluate $(-2m^2)^3$. Then evaluate.

_____

_____

4. Using the properties of exponents, demonstrate how to simplify $(a^2b^{-1})^0$.

_____

_____

5. Compare the expressions $-(rt)^2$ and $(-rt)^2$. Simplify each expression. How are the results related?

_____

_____

SELF-ASSESSMENT: Compare the process for calculating powers of numbers with the process for calculating powers of variables.

Alternative Assessment
All-You-Can Remember, Chapter 8, Form B

TASK: Solve a real-world problem by using the growth formula.

HOW YOU WILL BE SCORED: As you work through the task, your teacher will be looking for the following:

- how well you can identify and explain each of the variables in a growth or decay formula
- whether you can explain how to use a growth or decay formula to estimate exponential increases or decreases

Althea memorizes 50 Latin words and then forgets 5 of the words the next day. If Althea continues to forget the Latin words at the same rate of 10% per day, estimate how many words she will have forgotten after 7 days.

Use the formula $y = A(1 - r)^t$ for the following problems:

1. Describe what the variables A, r, t, and y represent for the given problem.

   _____

   _____

2. Assign the given values to the variables in the formula.

   _____

3. Make a table of the number of words that Althea remembers over a 10-day period.

4. Demonstrate how to use the data in your table to estimate the number of words that she will have forgotten after 7 days.

   _____

   _____

5. Explain why an exponential decay function was used to solve this problem.

   _____

   _____

SELF-ASSESSMENT: Explain how you could have solved this problem graphically.

_____

Quick Warm-Up: Assessing Prior Knowledge
9.1 Adding and Subtracting Polynomials

Simplify.

1. $-9b + 8b$ _____

2. $6y - y$ _____

3. $4m + (7 - m)$ _____

4. $c - (3c + 1)$ _____

5. $(3r + 4) + (-2r + 8)$ _____

6. $(-2g + 1) - (2g + 1)$ _____

7. $(5j - 7k) - (4j - 6k)$ _____

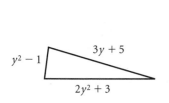

Lesson Quiz
9.1 Adding and Subtracting Polynomials

1. Name the polynomial $2y^3 + 8y - 3$ by the number of terms and

   by degree. _____

Simplify. Express all answers in standard form.

2. $(5x^2 + 2x - 3) + (3x - 5)$ _____

3. $(2x^3 - 3x^2 + 6x + 1) + (x^3 + 7x^2 - 9x - 6)$ _____

4. $(-3y^3 + 7y - 5y^2 + 2) + (8y^2 - 2y^3 - 3y - 9)$ _____

5. $(6t^4 - 7t^2 + 3) + (5t^3 + 2t^2)$ _____

6. $(2a^3 + 5a^2 + 7a + 2) - (9a^3 + 4a^2 + 2a + 6)$ _____

7. $(4p^3 + 3p - 1) - (7p^3 + 5p^2 + 3p + 1)$ _____

8. $(r^3 - 3r^2 - 7) - (r^2 - 14r + 4)$ _____

9. $(-5y^3 + 3y + 8) - (8y^3 + 7y^2 - 4y + 3)$ _____

10. Find the sum of the perimeters of the two triangles shown below. _____

```
       3y + 5
y² − 1  _____
       |       /           2y² + 1 | | 2y² + 1
       |_____/
       2y² + 3                      2y − 3
```

Quick Warm-Up: Assessing Prior Knowledge
9.2 Modeling Polynomial Multiplication

Simplify.

1. $(-7)(-2)$ _____

2. $(-3)(4t)$ _____

3. $3(s + 5)$ _____

4. $4(-n + 2)$ _____

5. $4 - (h + 2)$ _____

Lesson Quiz
9.2 Modeling Polynomial Multiplication

1. Which product of binomial factors is
   modeled at the right? _____

   a. $2(x + 1) = 2x + 2$

   b. $2x(x + 1) = 2x^2 + 2$

   c. $2x(x + 1) = 2x^2 + 2x$

Find each product by using the rules for special products.

2. $(x + 2)(x - 2)$ _____

3. $(y - 4)(y - 4)$ _____

4. $(2m + 3)(2m + 3)$ _____

5. $(d - 8)(d + 8)$ _____

6. $(3c - 1)(3c + 1)$ _____

7. $(5r + 4)(5r + 4)$ _____

Use the Distributive Property to find each product.

8. $2(x - 5)$ _____

9. $2r(r - 1)$ _____

10. $-5(g + 2)$ _____

11. $-s(s + 2)$ _____

Quick Warm-Up: Assessing Prior Knowledge
9.3 *Multiplying Binomials*

Simplify.

1. $(6z)(4z)$ _____

2. $-2(w + 5)$ _____

3. $6q - 9 - 5q$ _____

4. $(k^2 - 9k) + (k^2 + 9)$ _____

5. $(d^2 - 4d) + (4d - 1)$ _____

Lesson Quiz
9.3 *Multiplying Binomials*

1. Show how to find $(x + a)(x + b)$ by using the Distributive Property.

_____

2. Does $(x + 2)^2 = x^2 + 2^2$? Why or why not?

_____

Use the FOIL method to find each product.

3. $(x + 7)(x + 3)$ _____

4. $(y + 12)(y - 5)$ _____

5. $(r - 5)(r - 8)$ _____

6. $(a + 4)(a - 9)$ _____

7. $(2b + 3)(b - 1)$ _____

8. $(3t - 2)(t - 3)$ _____

9. $(p + 5)(2p - 2)$ _____

10. $\left(x - \frac{1}{2}\right)\left(x - \frac{1}{2}\right)$ _____

11. $\left(2y + \frac{1}{2}\right)\left(y - \frac{1}{2}\right)$ _____

12. $(3z - 2)(2z + 5)$ _____

A rectangle has a length of $y + 5$ units and a width of $y - 2$ units.

13. Draw a diagram in the space provided and label the dimensions.

14. Express the area of the rectangle in terms of y.

_____

Quick Warm-Up: Assessing Prior Knowledge
9.4 Polynomial Functions

Complete each function table.

1. $y = 2x + 1$

x	y
-3	
0	
1	
7	

2. $y = 7 - x$

x	y
-6	
-4	
2	
8	

Lesson Quiz
9.4 Polynomial Functions

A box is 15 centimeters long, 40 centimeters high and 6 centimeters wide.

1. Find the volume of the box. _____

2. Find the surface area of the box. _____

3. Find the volume of the box if the length is doubled. _____

4. Find the surface area of the box if the width is divided by 2. _____

5. Write a polynomial function that represents the
   volume of the box if the width is x centimeters. _____

**Use substitution to show that the functions in each pair are
equivalent for x-values of -2, -1, 0, 1, and 2.**

6. $y_1 = (x + 1)^2$ and $y_2 = x^2 + 2x + 1$ _____

7. $y_1 = (x - 2)(x + 3)$ and $y_2 = x^2 + x - 6$ _____

8. $y_1 = x^2 - 5x + 4$ and $y_2 = (x - 1)(x - 4)$ _____

**Use substitution to verify that each equation is an identity for
x-values of -2, -1, 0, 1, and 2.**

9. $(x + 5)(x + 2) = x^2 + 7x + 10$ _____

10. $x^2 - 5x - 14 = (x + 2)(x - 7)$ _____

Mid-Chapter Assessment
Chapter 9 (Lessons 9.1–9.4)

Write the letter that best answers the question or completes the statement.

_____ 1. Which expression is modeled by the algebra tiles?

a. $3x^2 - 2x + 5$
b. $3x^2 + 2x + 5$
c. $3x^2 - 2x - 5$
d. $3x^2 + 5$

_____ 2. What is the sum of $2x^3 - 3x^2 + 5x - 1$ and $5x^2 - 6x$?

a. $7x^3 - 9x^2 + 5x - 1$.
b. $2x^3 + 8x^2 + 11x + 1$.
c. $2x^3 + 2x^2 - x - 1$.
d. $7x^5 - 9x^3 + 5x - 1$.

_____ 3. When $x^3 - 2x + 5$ is subtracted from $2x^3 - 3x^2 - 2x + 1$, the result is

a. $x^3 - x^2 - 7x + 1$
b. $3x^3 - 3x^2 - 4x + 6$
c. $-x^3 + 3x^2 + 4$
d. $x^3 - 3x^2 - 4$

_____ 4. The product $3y^2(2y^2 - 1)$ is equal to

a. $5y^4 - 3y^2$.
b. $6y - 3y^2$.
c. $6y^4 - 3y^2$.
d. $5y^3 - 1$.

_____ 5. What is the factored form of $6n^3 - 3n^4$?

a. $3n^3(2 - n)$
b. $3n^4(2 - 1)$
c. $3n^2(2n + n^2)$
d. $3n^3(n - 2)$

_____ 6. The product of $(3a - 5)$ and $(2a + 3)$ is

a. $6a^2 - 19a - 15$.
b. $6a^2 + 19a - 15$.
c. $6a^2 - 15a + 9$.
d. $6a^2 - a - 15$.

Perform the indicated operations.

7. $(5x^4 - 2x^2 + 6x - 9) + (2x^4 + 5x^2 - 7x + 12)$ _____

8. $(4x^3 + 7x^2 - 5) - (2x^3 - 3x^2 + 5x - 9)$ _____

9. $3x^2(7x^2 - 5x + 2)$ _____

10. $(6x - 2)(2x + 1)$ _____

11. Write a function to represent the volume, V, of a cube with edges that are y centimeters long. _____

12. Use substitution to show that the equation $(x - 3)(x + 2) = x^2 - x - 6$ is true for x-values of $-2, -1, 0, 1,$ and 2. _____

Quick Warm-Up: Assessing Prior Knowledge
9.5 Common Factors

Find the greatest common factor (GCF) of each set of numbers.

1. 6, 15 _____   2. 16, 24, 60 _____   3. 3, 6, 14, 28 _____

Multiply.

4. $4(h^2 - 5)$ _____   5. $2b(b^2 - 9b)$ _____

Lesson Quiz
9.5 Common Factors

Find the GCF of each pair of numbers.

1. 25 and 75   _____   2. r^2t and $2r^3t^2$ _____

3. $3x^3y^2$ and $9xy^3$ _____   4. $7(a + b)$ and $5(a + b)$ _____

Factor each polynomial by using the GCF.

5. $3x^2 - 9$ _____   6. $5x^2 + 15x$ _____

7. $y^5 - y^3$ _____   8. $6t^4 + 18t^2 - 24t$ _____

9. $28 + 16n^3 - 48n$ _____   10. $3a^3b^2 - 15a^2b + 20ab^3$ _____

Write each polynomial as the product of two binomials.

11. $x(x - 3) + 5(x - 3)$ _____

12. $a(x - 2y) - b(x - 2y)$ _____

13. $(5 + m)2n + 3(5 + m)$ _____

14. $3(x - 1) + x(x - 1)$ _____

15. $pq(a - b)^n + st(a - b)^n$ _____

Factor by grouping.

16. $ab + ac - 5b - 5c$ _____

17. $12y^2z - 18y^3z^2 + 24y^5z^3 - 9$ _____

Quick Warm-Up: Assessing Prior Knowledge
9.6 Factoring Special Polynomials

Multiply.

1. $4(j + 6)$ _____

2. $(n - 9)(n + 9)$ _____

3. $(2t + 5)(2t + 5)$ _____

Factor.

4. $3k^2 + 21$ _____

5. $2y^2 - 15y$ _____

6. $6c^3 + 9c^2$ _____

Lesson Quiz
9.6 Factoring Special Polynomials

Find each product.

1. $(x + 2)^2$ _____

2. $(y - 3)(y + 3)$ _____

3. $(5a - 4)^2$ _____

4. $(2a - 5b)(2a + 5b)$ _____

Find the missing term in each perfect-square trinomial.

5. $x^2 + 10x +$ _____

6. $y^2 -$ _____ $+ 64$

7. $4a^2 +$ _____ $+ 9$

8. $9m^2 - 30m +$ _____

Factor each polynomial completely.

9. $y^2 - 81$ _____

10. $49a^2 - 25$ _____

11. $b^2 + 16b + 64$ _____

12. $m^2 - 6m + 9$ _____

13. $4x^2 - 16x + 16$ _____

14. $36p^2 + 36p + 9$ _____

15. $c^4 - d^4$ _____

16. $25x^4 - 4x^2$ _____

17. $9a^2 + 6ab + b^2$ _____

18. $y^2(y^2 - 9) - 4(y^2 - 9)$ _____

The area of a square is represented by $a^2 - 14a + 49$.

19. Express the length of each side in terms of a. _____

20. Express the perimeter of the square in terms of a. _____

Quick Warm-Up: Assessing Prior Knowledge
9.7 Factoring Quadratic Trinomials

Factor each polynomial by using the GCF.

1. $24v + 16$ _____   2. $6t^2 - 18t$ _____   3. $8a^4 + 16a^2 + 10$ _____

Factor each polynomial.

4. $n^2 - 144$ _____   5. $16z^2 - 40z + 25$ _____

Lesson Quiz
9.7 Factoring Quadratic Trinomials

1. Write all possible factor pairs of 24. _____

Write all the factor pairs of the third term of each polynomial. Then mark the pair that you would use to factor the given polynomial.

2. $x^2 - 3x - 10$ _____

3. $x^2 - 7x + 12$ _____

4. $x^2 + 11x + 18$ _____

5. $x^2 + 7x - 30$ _____

Factor each polynomial. In some cases you will need to factor out a constant monomial term.

6. $x^2 + 3x - 4$ _____   7. $2x^2 + 8x + 6$ _____

8. $x^2 - 6x + 5$ _____   9. $x^2 - x - 12$ _____

10. $y^2 - 36$ _____   11. $10x^2 - 5x$ _____

12. $9a^4 - 16a^2$ _____   13. $36b^2 - 12b^5$ _____

14. $d^2 + 7d - 18$ _____   15. $x^2 - 2x - 24$ _____

Quick Warm-Up: Assessing Prior Knowledge
9.8 *Solving Equations by Factoring*

Factor each polynomial.

1. $u^2 + 6u + 8$ _____

2. $k^2 + 4k + 4$ _____

3. $b^2 - b - 12$ _____

Solve each equation.

4. $8 - j = 18$ _____

5. $-3c = -42$ _____

6. $2(x + 10) = 0$ _____

Lesson Quiz
9.8 *Solving Equations by Factoring*

Identify the zeros of each function.

1. $y = (x + 3)(x - 1)$ _____

2. $y = (x - 5)(x + 5)$ _____

3. $y = (x + 4)(x + 1)$ _____

4. $y = x(x - 6)$ _____

5. $y = (2x + 4)(3x + 9)$ _____

6. $y = (3x - 15)^2$ _____

Solve by factoring.

7. $x^2 - x - 2 = 0$ _____

8. $x^2 + 4x + 4 = 0$ _____

9. $x^2 + x - 12 = 0$ _____

10. $x^2 - x - 30 = 0$ _____

11. $x^2 - 3x - 4 = 0$ _____

12. $2x^2 - 5x - 3 = 0$ _____

13. $2x^2 - 3x + 1 = 0$ _____

14. $4x^2 - 4 = 0$ _____

15. $3x^2 - x - 2 = 0$ _____

16. $4x^2 - 4x + 1 = 0$ _____

Chapter Assessment

Chapter 9, Form A, page 1

Write the letter that best answers the question or completes the statement.

_____ 1. The sum of $3r^4 - 7r^2 + r - 9$ and $r^4 + 5r^3 - 2r^2 + 1$ is

     a. $4r^4 - 12r^2 + 3r - 10$ b. $4r^4 - 2r^2 - r - 8$

     c. $4r^4 + 5r^3 - 5r^2 + r - 8$ d. $4r^4 + 5r^3 - 9r^2 + r - 8$

_____ 2. The difference when $x^3 - 5x^2 + 7$ is subtracted from $x^5 - 2x^3 + 3x^2 - 3$ is:

     a. $x^5 - x^3 - 3x^2 + 4.$ b. $x^5 - 3x^3 + 8x^2 - 10.$

     c. $-x^5 + x^3 - 8x^2 + 10.$ d. $x^5 - x^3 - 2x^2 + 4.$

_____ 3. Which expression is modeled by the algebra tiles?

     a. $4x^2 - 2x + 1$

     b. $(2x - 1)^2$

     c. $(2x - 1)(2x + 1)$

     d. $4x^2 + 2x - 1$

In Exercises 4–7, simplify each expression.

_____ 4. $6y^2(3y^2 - 1) =$

     a. $18y^4 - 6y^2$ b. $9y^4 - 6y$ c. $18y^4 - 1$ d. $18y^3 - 6y^2$

_____ 5. $(4p - 3)^2 =$

     a. $16p^2 - 6p + 9$ b. $16p^2 - 12p + 9$ c. $16p^2 - 24p + 9$ d. $16p^2 - 12p - 9$

_____ 6. $(5t - 3)(2t + 2) =$

     a. $10t^2 - 16t - 6$ b. $10t - t - 6$ c. $15t^2 + 4t + 4$ d. $10t^2 + 4t - 6$

_____ 7. $(3x - 2)(3x + 2) =$

     a. $9x^2 - 4$ b. $3x^2 - 6x - 4$ c. $9x^2 - 12x - 4$ d. $9x^2 - 2x - 4$

_____ 8. What is the GCF of $3a^4b^2c$ and a^2b^3c?

     a. $3a^2b^3c$ b. a^2b^2c c. $3a^4b^3c$ d. a^2b^3c

Chapter Assessment

Chapter 9, Form A, page 2

_____ 9. What is the factored form of $2a^2b^3 + 8a^3b^2 - 6a^2b$?

a. $2a^2b(b^2 + 4ab - 3)$ b. $2ab(ab^2 + 4ab - 3a)$
c. $2a^2b^3(8a^3b^2 - 6a^2b)$ d. $2a^2b^2(b + 4a - 3)$

_____ 10. What are the solutions to the equation $x^2 - 5x + 6 = 0$?

a. -2 and 3 b. 2 and -3 c. -2 and -3 d. 2 and 3

In Exercises 11–14, write the letter that represents the factored form of each expression.

_____ 11. $9x^2 - 36x + 36$

a. $9(x + 2)(x - 2)$ b. $3(x + 3)(3x + 4)$
c. $9(x - 4)(x + 1)$ d. $9(x - 2)^2$

_____ 12. $25a^2 - 9b^4$

a. $16(a + b^2)(a - b^2)$ b. $25a^2(1 - 3b^2)(1 + 3b^2)$
c. $(5a - 3b^2)(5a + 3b^2)$ d. $(5a - 3b^2)^2$

_____ 13. $6x^2 + x - 12$

a. $2(3x + 2)(x - 3)$ b. $(3x - 4)(2x + 3)$
c. $(3x + 4)(2x - 3)$ d. $6(x + 1)(x - 2)$

_____ 14. $2y(y - z)^2 - 5(y - z)^2$

a. $(2y - 5)(y - z)$ b. $(2y - 5)(y - z)^2$
c. $(2y - 5)(y + z)^2$ d. $(2y^2 - 5)(2y^2 + z)^2$

_____ 15. Which expression represents the shaded region in the figure at right?

a. $x^2 - 25$ b. $(x - 5)^2$
c. $x^2 + 25$ d. $(x + 5)^2$

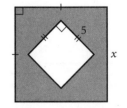

_____ 16. The perimeter of a rectangle with a length of $y + 5$ units and a width of $y - 2$ units can be represented by

a. $y^2 + 3y - 10$.   b. $2y + 3$.   c. $4y + 6$.   d. $y^2 + 7y - 10$.

Chapter Assessment

Chapter 9, Form B, page 1

1. Identify $9x^3 + 1$ by degree and by the number of terms. _____

2. Explain why $3x + 4y$ is considered prime.

    _____

3. Give an example of a perfect-square trinomial.

    _____

4. If you wrote $x^2 - 8x + 12$ in factored form, would the signs in the monomial factors be the same or opposite? Why?

    _____

5. If you wrote $x^2 - 8x + 12$ in factored form, would the signs in the monomial factors be positive or negative? Why?

    _____

Add. Express all answers in standard form.

6. $3y^3 - 8y^2 + 7y - 9$ and $9y^3 + 7y^2 - 5y - 6$ _____

7. $w^2 + 12w - 6w^3 + 3$ and $5w^3 - w^2 + 3w - 11$ _____

8. $5a^4 - 9a^2 + 7a + 15$ and $a^4 + 7a^2 - 2a^3 - 8$ _____

Subtract. Express all answers in standard form.

9. $x^3 + 3x^2 - 5x$ from $4x^3 - 9x^2 + 12$ _____

10. $6b^3 - 2b^2 + 7$ from $5b^3 - 11b^2 + 3b - 6$ _____

11. $8d^4 + 5d - 3d^2 + 1$ from $13d^4 - 5d^2 + 3d^3 - 1$ _____

Simplify. Express all answers in standard form.

12. $(3 - 2x + x^3) + (3x - 5x^3 + 8)$ _____

13. $(7 - 9y^2 + 3y^3) - (3y^2 + 8 - 9y^3)$ _____

14. Express the area of the triangle shown at right in terms of x.

    _____

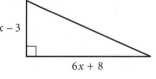

Algebra 1

Chapter Assessment
Chapter 9, Form B, page 2

Find each product.

15. $3w(5w^2 - 1)$ _____

16. $(4a - 2b^2)(4a + 2b^2)$ _____

17. $(x + 3)^2$ _____

18. $(2x - 3)^2$ _____

19. $(d + 5)(d - 3)$ _____

20. $(3t - 8)(2t + 3)$ _____

Factor each polynomial completely.

21. $9r^2 - 4$ _____

22. $z^2 + 7z - 18$ _____

23. $m^2 - 10m + 25$ _____

24. $7a^3 + 28a^2 - 35a$ _____

25. $y^2 - 3y - 18$ _____

26. $b(b - 1) + 5(b - 1)$ _____

Solve each equation.

27. $x^2 + 3x - 4 = 0$ _____

28. $x^2 - 4 = 0$ _____

29. $x^2 - 8x + 15 = 0$ _____

30. $x^2 - 2x + 1 = 0$ _____

31. Express the perimeter of the rectangle shown at right in terms of *x*.

_____

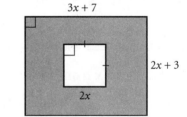

32. Express the area of the rectangle shown at right in terms of *x*.

_____

33. Express the area of the shaded region in terms of *x*. _____

The area of a square is represented by $4p^2 - 12p + 9$.

34. Express the length of each side in terms of *p*. _____

35. Express the perimeter of the square in terms of *p*. _____

Alternative Assessment
Exploring Polynomials, Chapter 9, Form A

TASK: Explore addition, subtraction, multiplication, and division of polynomials.

HOW YOU WILL BE SCORED: As you work through the task, your teacher will be looking for the following:

- whether you can explain and model with tiles the process of adding and subtracting polynomials
- how well you can explain and model with tiles the process of multiplying and dividing polynomials

1. Explain how to subtract $3x^2 + 1$ from $4x^2 + 2x - 1$. Then find the difference.

_____

_____

2. Model the subtraction of $3x^2 + 1$ from $4x^2 + 2x + 1$ by using algebra tiles.

3. Explain how to find $(x + 1)(x - 6)$. Then find the product.

_____

4. Model the product of $(x + 1)(x - 6)$ by using algebra tiles.

5. Write $2x^4 - 8x^2$ in factored form.

_____

SELF-ASSESSMENT: Use algebra tiles to model the addition, subtraction, multiplication, or division of two binomials.

Alternative Assessment

Factoring Polynomials, Chapter 9, Form B

TASK: Factor polynomials by using the greatest common factor and by recognizing special products.

HOW YOU WILL BE SCORED: As you work through the task, your teacher will be looking for the following:

- how well you understand the concept of factoring
- how effectively you can communicate your responses in writing
- whether you can write generalizations

1. Describe the method you would use to determine whether $3x - 1$ is a factor of $3x^2 - 4x + 1$. If $3x - 1$ is a factor, how would you find the remaining factor of $3x^2 - 4x - 1$?

_____

_____

Factor each polynomial. Which polynomials, if any, are prime?

2. $2x^3 - 4x^2 + 8x$ 3. $6a - 5b$ 4. $10xy + y^2$

_____   _____   _____

5. Write a generalization about how to determine whether a polynomial is prime or composite.

_____

_____

Write each polynomial as a product of its factors.

6. $5n^2 + 4n - 1$ 7. $100x^2 - 25$ 8. $9p^2 + 6p + 1$

_____   _____   _____

9. Describe how to check the factors in Exercises 6–9. Then check your results.

_____

_____

SELF-ASSESSMENT: Use technology to graph a polynomial function. Then use the graph to find the zeros of the function. Use the zeros to determine the factors of the polynomial that defines the function. Explain your methods.

Quick Warm-Up: Assessing Prior Knowledge
10.1 *Graphing Parabolas*

Solve by factoring.

1. $x^2 + 4x - 12 = 0$ _____

2. $x^2 - x - 30 = 0$ _____

3. $x^2 - 8x + 16 = 0$ _____

4. $x^2 - 1 = 0$ _____

5. $x^2 - x = 0$ _____

Lesson Quiz
10.1 *Graphing Parabolas*

Compare the graphs of the following functions to the graph of
$y = x^2$. Describe the vertical and horizontal translations of each
vertex.

1. $y = (x - 4)^2 + 1$ _____

2. $y = \frac{1}{2}x^2 - 3$ _____

3. $y = 3(x + 2)^2$ _____

4. $y = -(x - 1)^2 + 2$ _____

Find the vertex and axis of symmetry for the graph of each function.

5. $y = (x + 3)^2 - 1$ _____

6. $y = (x - 5)^2$ _____

Sketch a graph of each function.

7. $y = (x - 2)^2 - 2$

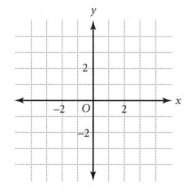

8. $y = -2(x + 3)^2 + 1$

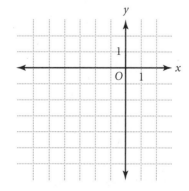

Quick Warm-Up: Assessing Prior Knowledge
10.2 Solving Equations of the Form ax² = k

Find each product.

1. $(-5)(-5)$ _____

2. $(11)(11)$ _____

Solve each equation.

3. $2x - 5 = 11$ _____

4. $-3y + 12 = 24$ _____

5. $-16t + 320 = 0$ _____

Lesson Quiz
10.2 Solving Equations of the Form ax² = k

1. Complete the table of values for the function $f(t) = -8t^2 + 100$.

t	0	1	2	3	4
$f(t)$	100	92			

2. Approximately where does the function cross the x-axis? _____

Find each positive square root. Round answers to the nearest hundredth when necessary.

3. $\sqrt{169}$ _____

4. $\sqrt{81}$ _____

5. $\sqrt{225}$ _____

6. $\sqrt{17}$ _____

Solve each equation. Round answers to the nearest hundredth when necessary.

7. $x^2 = 49$ _____

8. $x^2 = \dfrac{100}{225}$ _____

9. $(x + 2)^2 - 16 = 0$ _____

10. $(x - 5)^2 - 36 = 0$ _____

11. $(x - 3)^2 - 7 = 0$ _____

12. $(x + 4)^2 = 25$ _____

For the function $y = (x + 5)^2 - 4$,

13. find the vertex. _____

14. find the axis of symmetry. _____

15. find the zeros. _____

16. sketch a graph on the grid provided.

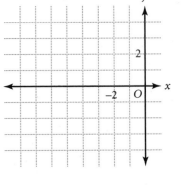

Quick Warm-Up: Assessing Prior Knowledge
10.3 Completing the Square

Evaluate each expression.

1. $\left(\frac{6}{2}\right)^2$ _____

2. $\left(-\frac{10}{2}\right)^2$ _____

3. $\left(\frac{7}{2}\right)^2$ _____

4. $\left(-\frac{1}{2}\right)^2$ _____

Factor each expression.

5. $x^2 + 2x + 1$ _____

6. $x^2 + 8x + 16$ _____

7. $x^2 - 10x + 25$ _____

Lesson Quiz
10.3 Completing the Square

1. What is the vertex for the graph of an equation of the form

$y = (x - h)^2 + k$? _____

Complete the square for each expression. Then write each expression in factored form.

2. $x^2 + 8x$ _____

3. $x^2 - 4x$ _____

4. $x^2 - 14x$ _____

Rewrite each function in the form $y = (x - h)^2 + k$.

5. $y = x^2 + 12x + 36 - 36$ _____

6. $y = x^2 - 18x + 81 - 81$ _____

Refer to the function $y = x^2 + 16x$ to complete Exercises 7–9.

7. Complete the square and rewrite the equation in the form

$y = (x - h)^2 + k.$ _____

8. Find the vertex. _____

9. Find the maximum or minimum value. _____

Rewrite each function in the form $y = (x - h)^2 + k$. Find each vertex.

10. $y = x^2 - 4x + 3$ _____

11. $y = x^2 + 3x + 5$ _____

Mid-Chapter Assessment

Chapter 10 (Lessons 10.1–10.3)

Write the letter that best answers the question or completes the statement.

_____ 1. What is the vertex of the graph of $y = \frac{1}{2}(x - 1)^2 + 3$?

     a. $\left(-\frac{1}{2}, 3\right)$ **b.** $(-1, 3)$ **c.** $\left(\frac{1}{2}, -3\right)$ **d.** $(1, 3)$

_____ 2. What is the axis of symmetry of the graph of $y = -3(x + 1)^2 - 2$?

     a. $x = 1$ **b.** $x = -1$ **c.** $x = 2$ **d.** $x = -2$

_____ 3. The zeros of $x^2 + 5x - 24$ are

     a. 5 and 3.   **b.** -5 and 4.   **c.** 3 and -8.   **d.** -3 and 8.

_____ 4. What are the values of x if $(x - 3)^2 + 4 = 20$?

     a. $x = 7, x = -1$ **b.** $x = 3, x = -4$ **c.** $x = 3 \pm 2\sqrt{6}$ **d.** $x = -2 \pm 2\sqrt{6}$

_____ 5. Which of the following completes the square for $x^2 + 5x$?

     a. $\left(\frac{5}{2}\right)^2$ **b.** 5^2 **c.** $\left(\frac{25}{4}\right)^2$ **d.** $-\left(\frac{5}{2}\right)^2$

Solve each equation. Round answers to the nearest hundredth when necessary.

6. $x^2 = 225$ _____

7. $(x - 1)^2 = 36$ _____

8. $(x + 2)^2 - 5 = 0$ _____

9. $(x - 5)^2 - 1 = 0$ _____

For the function $y = x^2 - 6x + 1$:

10. Rewrite the equation in the form $y = (x - h)^2 + k$.

_____

11. Find the vertex. _____

12. Find the axis of symmetry. _____

13. Find the zeros. _____

14. Sketch the graph of the function on the grid provided.

Quick Warm-Up: Assessing Prior Knowledge

10.4 Solving Equations of the Form $x^2 + bx + c = 0$

Solve each equation.

1. $(x - 3)^2 - 16 = 0$ _____

2. $(x + 4)^2 - 25 = 0$ _____

Factor each expression.

3. $x^2 + x - 6$ _____

4. $x^2 - 8x + 12$ _____

Lesson Quiz

10.4 Solving Equations of the Form $x^2 + bx + c = 0$

Solve each equation by factoring.

1. $x^2 - 10x + 24 = 0$ _____

2. $x^2 + 11x + 30 = 0$ _____

3. $x^2 - 7x - 18 = 0$ _____

4. $x^2 + 4x - 21 = 0$ _____

Solve by completing the square.

5. $x^2 + 6x = 0$ _____

6. $x^2 - 4x - 1 = 0$ _____

7. $x^2 - 10x + 5 = 0$ _____

8. $x^2 + 2x - 7 = 0$ _____

9. Graph the function $f(x) = x^2 - 6x + 5$ on the grid provided.

Use the graph from Exercise 9 to find the values of *x* for each of the following values of *y*. Check by substitution.

10. $y = 12$ _____

11. $y = -4$ _____

12. $y = 5$ _____

13. $y = 0$ _____

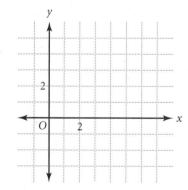

Quick Warm-Up: Assessing Prior Knowledge
10.5 The Quadratic Formula

Find each square root.

1. $\sqrt{64}$ _____

2. $\pm\sqrt{144}$ _____

Evaluate each expression.

3. $6^2 - 4(1)(3)$ _____

4. $2^2 - 4(1)(-3)$ _____

5. $\dfrac{-6 + \sqrt{6^2 - 4(1)(5)}}{2(1)}$ _____

Lesson Quiz
10.5 The Quadratic Formula

Identify *a, b,* and *c* for each quadratic equation.

1. $x^2 - 7x + 11 = 0$ _____

2. $2x + 3x^2 - 8 = 0$ _____

Find the value of the discriminant and determine the number of real solutions for each equation.

3. $x^2 - 3x + 8 = 0$ _____

4. $x^2 + 12x - 13$ _____

5. $2x^2 - 7x + 3 = 0$ _____

Use the quadratic formula to solve each equation. Round your answers to the nearest hundredth when necessary. Check by substitution.

6. $x^2 + 7x + 6 = 0$ _____

7. $2y^2 + 7y + 3 = 0$ _____

8. $-x^2 + 5x + 6 = 0$ _____

9. $-r^2 - 6r + 3 = 0$ _____

Choose any method to solve each quadratic equation. Round your answers to the nearest hundredth when necessary.

10. $y^2 + 8y + 15 = 0$ _____

11. $y^2 - 25 = 0$ _____

12. $x^2 - 6x + 1 = 0$ _____

13. $3t^2 - 18t = 0$ _____

Quick Warm-Up: Assessing Prior Knowledge
10.6 *Graphing Quadratic Inequalities*

Solve each equation.

1. $x^2 - 2x - 15 = 0$ _____

2. $x^2 + 3x - 28 = 0$ _____

Tell whether the given ordered pair satisfies the given inequality.

3. $(1, 4); y > -2x + 3$ _____

4. $(-2, 3); y \le -3x + 4$ _____

5. $(-1, -2); 5x - y \ge 7$ _____

Lesson Quiz
10.6 *Graphing Quadratic Inequalities*

Solve each quadratic inequality by using the Zero Product Property. Graph each solution on the number line provided.

1. $x^2 + 7x + 10 > 0$

    <----+----+----+----+----+----+----+----+----+----+----+----+----> x
            −6   −4   −2   0   2   4   6

2. $x^2 - x - 30 < 0$

    <----+----+----+----+----+----+----+----+----+----+----+----+----> x
            −6   −4   −2   0   2   4   6

3. $x^2 + 4x - 12 \ge 0$

    <----+----+----+----+----+----+----+----+----+----+----+----+----> x
            −6   −4   −2   0   2   4   6

4. $x^2 + x - 20 \le 0$

    <----+----+----+----+----+----+----+----+----+----+----+----+----> x
            −6   −4   −2   0   2   4   6

Graph each quadratic inequality on the grid provided. Shade the solution region.

5. $y \ge x^2 - 2$

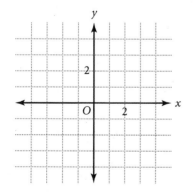

6. $y < \frac{1}{2}x^2 + 1$

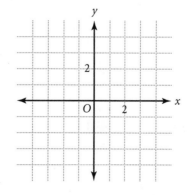

Chapter Assessment

Chapter 10, Form A, page 1

Write the letter that best answers the question or completes the statement.

 1. What is the vertex of the graph of $y = \frac{1}{2}(x + 6)^2 - 2$?

  a. $(3, -2)$ b. $(-3, -1)$ c. $(6, -2)$ d. $(-6, -2)$

_____ 2. The axis of symmetry of the graph of $y = -(x - 3)^2 + 5$ is

  a. $x = 3$.   b. $x = -3$.   c. $x = 5$.   d. $x = -5$.

_____ 3. What is the factored form of $x^2 - 3x - 28$?

  a. $(x + 7)(x - 4)$ b. $(x - 3)(x + 28)$

  c. $(x - 7)(x + 4)$ d. $(x - 14)(x + 2)$

 4. The solutions to $(x + 3)^2 - 25 = 0$ are

  a. -3 and 25.   b. -3 and 5.

  c. -8 and 2.   d. 8 and -2.

_____ 5. Which of the following shows $x^2 - 6x + 5 = 0$ in the form $(x - h)^2 + k = 0$?

  a. $(x - 6)^2 + 5 = 0$ b. $(x - 3)^2 + (-4)$

  c. $(x - 3)^2 + 11 = 0$ d. $(x - 6)^2 = 0$

_____ 6. The graph of $y = x^2 - 2x - 3$ intersects the x-axis at

  a. $(3, 0)$ and $(-1, 0)$.   b. $(0, 3)$ and $(0, -1)$.

  c. $(-3, 0)$ and $(1, 0)$.   d. $(3, 0)$ and $(1, 0)$.

 7. The solutions to $x^2 - 6x - 27 = 0$ are

  a. -9 and 3.   b. 6 and -3.   c. -6 and 3.   d. -3 and 9.

_____ 8. The zeros of $y = x^2 + 8x + 7$ are

  a. 1 and 7.   b. 7 and 8.

  c. -7 and -1.   d. -7 and -8.

 9. Which of the following would complete the square for $x^2 + x$?

  a. $\frac{1}{2}$.   b. $-\frac{1}{2}$.   c. $\frac{1}{4}$.   d. $-\frac{1}{4}$.

_____ 10. For $f(x) = x^2 - 4x + 3$, the values of x where $f(x) = 3$ are

     a. 1 and 3.   b. -4.   c. 4 and 0.   d. 2 and 3.

_____ 11. What is the discriminant of $2x^2 - 3x - 5 = 0$?

     a. 7   b. 31   c. 36   d. 49

_____ 12. What is the value of $-b$ for $7 + 3x^2 - 8x = 0$?

     a. 3   b. -8 c. 8   d. 64

_____ 13. Using the quadratic formula, the solutions of $3y^2 + 11y = 4$ are

     a. $\frac{1}{3}$ and -4.   b. $-\frac{1}{3}$ and 4.   c. $\frac{2}{3}$ and -4.   d. -3 and 4.

_____ 14. Which of the following shows the solution to $x^2 - 6x + 8 \leq 0$?

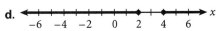

_____ 15. Which inequality is graphed at right?

     a. $y \geq x^2 + 4x + 3$

     b. $y > x^2 + 4x + 3$

     c. $y < x^2 + 4x + 3$

     d. $y \leq x^2 + 4x + 3$

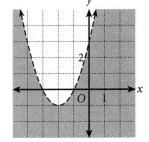

_____ 16. The profit function $p = -x^2 + 18x - 56$ describes a relationship between x, the number of calculators produced (in thousands), and p, the profit (in thousands of dollars). What is the break-even point for the company?

     a. between 4000 and 14,000

     b. less than 4000 or greater than 14,000

     c. 4000 or 14,000

     d. 14,000 only

Chapter Assessment

Chapter 10, Form B, page 1

Find the vertex and axis of symmetry for each function.

1. $y = \frac{1}{2}(x + 1)^2 - 3$ _____

2. $y = -(x - 4)^2 - 5$ _____

Factor each polynomial completely.

3. $x^2 - 8x + 15$ _____

4. $x^2 + 4x + 3$ _____

5. $x^2 + 5x - 24$ _____

6. $3x^2 - 9x + 6$ _____

The graph below represents the relationship between the time, in seconds, after a projectile is propelled vertically into the air and the height, in feet, that it reaches.

7. What is the maximum height reached by the projectile?

_____

8. How long does it take the projectile to reach its maximum height?

_____

9. From the time the projectile has been fired, how long does it take the projectile to hit the ground?

_____

10. What is the axis of symmetry of the graph?

_____

Find each square root. Round answers to the nearest hundredth when necessary.

11. $\sqrt{225}$ _____

12. $\sqrt{121}$ _____

13. $\sqrt{23}$ _____

14. $\sqrt{95}$ _____

Solve each equation. Round answers to the nearest hundredth when necessary.

15. $x^2 = 196$ _____

16. $(x - 3)^2 = 25$ _____

17. $(x + 5)^2 - 1 = 0$ _____

18. $(x - 2)^2 = 17$ _____

Chapter Assessment

Chapter 10, Form B, page 2

Write each function in the form $y = (x - h)^2 + k$. Find each vertex.

19. $y = x^2 + 6x - 9$ _____

20. $y = x^2 - 5x + \dfrac{3}{4}$ _____

Choose any method to solve each equation.

21. $x^2 - 8x - 20 = 0$ _____

22. $t^2 + 6t = 0$ _____

23. $w^2 - 25 = 0$ _____

24. $2y^2 - 11y + 15 = 0$ ____

25. $3x^2 - 7x - 5 = 0$ _____

26. $5r^2 + 13r - 3 = 0$ _____

Find the points of intersection for the graphs of each pair of equations.

27. $y = 2x - 2$ and $y = x^2 - 2x + 1$

28. $y = x + 5$ and $y = x^2 + 6x - 9$

_____

_____

Solve each quadratic inequality. Graph the solution on the number line provided.

29. $x^2 + 3x - 10 \geq 0$

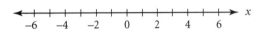

30. $x^2 - 5x + 4 < 0$

Graph each quadratic inequality.

31. $y > x^2 + 4x$

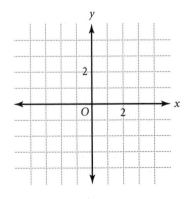

32. $y \leq x^2 - 2x + 3$

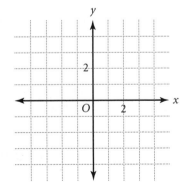

Alternative Assessment

Examining a Quadratic Function, Chapter 10, Form A

TASK: Explain the meaning of the vertex, the axis of symmetry, and the zeros of quadratic functions in a real-world problem.

HOW YOU WILL BE SCORED: As you work through the task, your teacher will be looking for the following:

- whether you can determine the vertex, the axis of symmetry, and the zeros of a quadratic function
- whether you can find the maximum or minimum of a quadratic function from its vertex

Carmen sells an average of 20 teddy bears per day. She charges $10 per bear. A survey indicates that she could sell 5 more teddy bears each day for each $1 decrease in the price. The table below shows the results of the survey.

Let x represent the number of $1 decreases in the price of a teddy bear and y represent the total income from sales.

x	0	1	2	3	4	5	6	7	8	9	10
y	200	225	240	245	240	225	200	165	125	65	0

1. Calculate the second differences. Then explain how the second differences indicate that the values of this function represent a quadratic polynomial.

_____

2. Work backward from the pattern to find the values of x when y equals 0. Then interpret this information in terms of the price Carmen charges for a teddy bear.

_____

3. Describe how to determine the axis of symmetry for this function from the x-value you found in Exercise 2.

_____

4. Explain how to determine the vertex of this function. What is the vertex? What is the maximum or minimum value?

_____

SELF-ASSESSMENT: Explain how to use the function $y = -5x^2 + 30x + 200$ to find the price Carmen should charge to receive the maximum income.

Alternative Assessment

Exploring Quadratic Equations, Chapter 10, Form B

TASK: Solve quadratic equations by using different methods.

HOW YOU WILL BE SCORED: As you work through the task, your teacher will be looking for the following:

- how well you can describe the steps for finding the solutions to a quadratic equation by completing the square
- whether you can use the quadratic formula to a find the solutions to a quadratic equation

1. Describe how you would find the solutions to $n^2 + 2n - 24 = 0$ by completing the square. Then find the solutions by completing the square.

_____

_____

2. Solve $2x^2 - x - 15 = 0$ by factoring. Describe how you can check your solutions.

_____

_____

3. Describe how you would use the quadratic formula to solve $x^2 = 3x + 4$. Then use the quadratic formula to find the solutions.

_____

_____

4. How would you decide whether to use factoring, completing the square, or the quadratic formula to find the solutions to $4t^2 + 12t = 7$? Use your chosen method to solve the equation.

_____

_____

_____

_____

SELF-ASSESSMENT: Describe any concepts that gave you difficulty as you worked through the task, and explain how you worked through the difficulties.

Quick Warm-Up: Assessing Prior Knowledge
11.1 *Inverse Variation*

In each case, *y* varies directly as *x*.

1. If *y* is 18 when *x* is 3, find *y* when *x* is 36. _____

2. If *y* is 24 when *x* is 5, find *y* when *x* is 6. _____

3. If *y* is 45 when *x* is 9, find *x* when *y* is 90. _____

4. If *y* is 10 when *x* is 12, find *x* when *y* is 42. _____

5. If *y* is 3.8 when *x* is 4, find *x* when *y* is 13.3. _____

Lesson Quiz
11.1 *Inverse Variation*

Determine whether each equation represents an inverse variation.

1. $\frac{d}{t} = 30$ _____   2. $\frac{8}{a} = \frac{b}{10}$ _____   3. $x \cdot y = 5$ _____   4. $\frac{1}{x} = 3$ _____

For Exercises 5–10, *y* varies inversely as *x*,

5. if *y* is 10 when *x* is 5, find *x* when *y* is 25. _____

6. if *y* is 7 when *x* is 12, find *x* when *y* is −6. _____

7. if *x* is $\frac{1}{2}$ when *y* is 16, find *y* when *x* is 32. _____

8. if *x* is −12 when *y* is $\frac{3}{4}$, find *y* when *x* is −3. _____

9. if *y* is 2 when *x* is 1000, find *x* when *y* is 1000. _____

10. if *x* is $\frac{2}{3}$ when *y* is $\frac{1}{2}$, find *y* when *x* is $\frac{1}{3}$. _____

11. The speed of a gear, in revolutions per minute, is inversely proportional to the number of the gear's teeth. If a gear with 15 teeth revolves at a speed of 700 revolutions per minute, at what speed should a gear with 21 teeth revolve? _____

12. Time varies inversely as the average speed over a given distance. When Tony travels 5 hours, his average speed is 50 miles per hour. Find Tony's time when his average speed is 40 miles per hour. _____

Algebra 1

NAME _____ CLASS _____ DATE _____

Quick Warm-Up: Assessing Prior Knowledge
11.2 *Rational Expressions and Functions*

Solve each equation.

1. $4y - 15 = 0$ _____

2. $5c^2 - 20c = 0$ _____

3. $m^2 - 2m + 1 = 0$ _____

4. $a^2 - 7a - 18 = 0$ _____

Lesson Quiz
11.2 *Rational Expressions and Functions*

1. Express $f(x) = \dfrac{3x + 5}{x}$ in a form which shows that it is a

   transformation of the parent function $g(x) = \dfrac{1}{x}$. _____

For what values of the variable are these rational expressions undefined?

2. $\dfrac{5x - 2}{x}$ _____

3. $\dfrac{4}{x + 1}$ _____

4. $\dfrac{2x + 3}{x^2 - 2x}$ _____

5. $\dfrac{x^2 - 3x + 2}{x^2 - x - 6}$ _____

Evaluate each rational expression for *x* = 2 and *x* = −1. Write
***undefined* if appropriate.**

6. $\dfrac{2x + 2}{x}$ _____

7. $\dfrac{3x}{x + 1}$ _____

8. $\dfrac{x^2 - 2}{x^2 - 6}$ _____

9. $\dfrac{x^2 + 3x}{x^2 - x + 3}$ _____

Graph each rational function. List any values of *x* for which the
function is undefined.

10. $f(x) = \dfrac{1}{x - 2}$

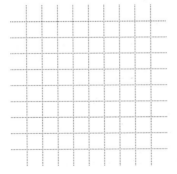

11. $f(x) = \dfrac{1}{x + 2} - 1$

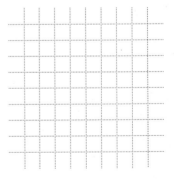

_____   _____

NAME _____ CLASS _____ DATE _____

Quick Warm-Up: Assessing Prior Knowledge
11.3 *Simplifying Rational Expressions*

Simplify. Write answers with positive exponents only.

1. $\dfrac{w^4}{w}$ _____

2. $\dfrac{3x^2y^5}{15xy^2}$ _____

3. $\dfrac{a}{a^7}$ _____

4. $\dfrac{24m^6n^4}{8m^2n^7}$ _____

Factor completely.

5. $5k - 35$ _____

6. $r^2 + 2r - 24$ _____

7. $x^2 - 1$ _____

8. $3b^2 - 42b + 72$ _____

Lesson Quiz
11.3 *Simplifying Rational Expressions*

1. Write 60 as the product of prime factors. _____

2. Factor $x^2 - 3x - 10$. _____

3. How can $5 - x$ be written in terms of $x - 5$? _____

Write the common factors of the numerator and the denominator of each rational expression.

4. $\dfrac{24}{60}$ _____

5. $\dfrac{3a^2b^3}{18a^2b}$ _____

6. $\dfrac{2d^2 - 5d}{d^2}$ _____

7. $\dfrac{3m^2 - 6m}{m^2 - 2m}$ _____

8. $\dfrac{(x + 3)^2}{x^2 + 2x - 3}$ _____

9. $\dfrac{y^2 + 3y - 10}{y^2 - 5y + 6}$ _____

Simplify each expression and state any restrictions on the variables.

10. $\dfrac{12z^2}{72z^5}$ _____

11. $\dfrac{48p^3q}{16pq^2}$ _____

12. $\dfrac{4x^2 - 9}{4x^2 + 6x}$ _____

13. $\dfrac{2x^2 - 8x}{3x - 12}$ _____

14. $\dfrac{12 - 3a}{a^2 - 16}$ _____

15. $\dfrac{(m + 1)^2}{m^2 - 1}$ _____

16. $\dfrac{y^2 + y - 12}{2y^2 + 8y}$ _____

17. $\dfrac{10b - 5b^2}{b^2 + 3b - 10}$ _____

18. The area of a certain rectangle with a width of $x + 1$ is represented by $x^2 - x - 2$. Find the rectangle's length.

_____

Quick Warm-Up: Assessing Prior Knowledge
11.4 Operations With Rational Expressions

Evaluate. Write answers in simplest form.

1. $\dfrac{7}{12} + \dfrac{1}{12}$ _____

2. $\dfrac{1}{2} + \dfrac{2}{3}$ _____

3. $\dfrac{9}{10} - \dfrac{3}{4}$ _____

4. $\dfrac{2}{5} \cdot \dfrac{5}{6} \cdot \dfrac{4}{5}$ _____

5. $\dfrac{3}{8} \div \dfrac{2}{3}$ _____

6. $\dfrac{3}{5} \div 6$ _____

Lesson Quiz
11.4 Operations With Rational Expressions

Answer each question about the fractions $\dfrac{5x^2}{2y}$ and $\dfrac{2y}{3x}$.

1. What is the common denominator found in Exercise 1? _____

2. How can you write $\dfrac{5x^2}{2y}$ with the common denominator found

   in Exercise 1? _____

3. When multiplying these expressions, how can you simplify the
   multiplication process?

   _____

**Perform the indicated operations. Simplify, and state the
restrictions on the variable.**

4. $\dfrac{5}{2a} + \dfrac{3}{5a}$ _____

5. $\dfrac{2}{3b^2} - \dfrac{5}{2b}$ _____

6. $\dfrac{6x}{5y^2} \cdot \dfrac{2y}{3x^3}$ _____

7. $\dfrac{3}{m^2} \div \dfrac{15}{m}$ _____

8. $\dfrac{r}{r-2} + 3$ _____

9. $\dfrac{9}{t-3} - \dfrac{7}{3-t}$ _____

10. $\dfrac{3n-6}{5} \cdot \dfrac{15}{n-2}$ _____

11. $\dfrac{p^2}{p^2-25} \div \dfrac{p}{p+5}$ _____

12. $\dfrac{x^2-3x}{x^2-2x-3}$ _____

13. $\dfrac{1}{4w-4} - \dfrac{1}{2w^2}$ _____

14. $\dfrac{z+3}{z-3} \cdot \dfrac{z^2-9}{4z+12}$ _____

15. $\dfrac{3-3d}{d^2-1} \div \dfrac{1}{d+1}$ _____

Mid-Chapter Assessment
Chapter 11 (Lessons 11.1–11.4)

Write the letter that best answers the question or completes the statement.

_____ 1. For what value(s) of a is $f(a) = \dfrac{a + 3}{a^2 - 6a + 5}$ undefined?

      a. -5 and -1 **b.** -3 **c.** 5 and 1   **d.** 0

_____ 2. Which function is graphed below?

      a. $y = \dfrac{1}{x} + 2$

      b. $y = \dfrac{1}{x} - 2$

      c. $y = -\dfrac{1}{x + 2}$

      d. $y = -\dfrac{1}{x} + 2$

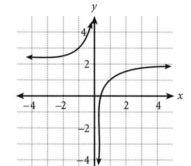

_____ 3. Given that y varies inversely as x, if y is 6 when x is 8, what is the value of x when y is 12?

      a. 2   **b.** 4   **c.** 24   **d.** 48

_____ 4. What is the common factor of the numerator and the denominator of $\dfrac{3x + 12}{x^2 + 3x - 4}$?

      a. $x + 4$ **b.** $3x$ **c.** $x + 3$ **d.** $3x(x + 4)$

Simplify.

5. $\dfrac{9a^3b^3}{36a^2b^5}$ _____

6. $\dfrac{12(x^2 - y^2)}{4(x - y)^2}$ _____

7. $\dfrac{2m^2 - 6m}{m^2 - 9}$ _____

8. $\dfrac{p^2 + 2p - 8}{p^2 + p - 6}$ _____

Perform the indicated operations. Simplify, and state restrictions on the variables.

9. $\dfrac{-2}{3a^2b} + \dfrac{5}{ab^2}$ _____

10. $\dfrac{8}{x - 2} - \dfrac{1}{2 - x}$ _____

11. $\dfrac{5r^2 + 15r}{r^2 + 4r + 3} \cdot \dfrac{r^2 - 1}{10r}$ _____

12. $\dfrac{3w}{4w - 12} \div \dfrac{12w^2}{(w - 3)^2}$ _____

Quick Warm-Up: Assessing Prior Knowledge
11.5 Solving Rational Equations

Solve each equation.

1. $9g - 16 = 38$ _____

2. $\frac{k}{4} + 9 = 3$ _____

3. $2n^2 - 12n + 18 = 0$ _____

4. $t^2 + 3t = 10$ _____

5. Simplify $\frac{x^2 - 4}{6x} \cdot \frac{x}{x + 2}$. State any restrictions on the variable _____

Lesson Quiz
11.5 Solving Rational Equations

1. What two equations would you graph to solve $\frac{2}{3}x + \frac{1}{x} = 5$? _____

Solve each rational equation by using the least common denominator.

2. $\frac{4x - 3}{4} + \frac{2x}{3} = 6$ _____

3. $\frac{5}{2x} - \frac{2}{5x} = \frac{3}{10}$ _____

4. $\frac{x}{x + 3} + \frac{5}{x - 3} = 1$ _____

5. $\frac{5x}{x^2 + 2x - 3} + \frac{7}{x - 1} = 1$ _____

Solve each rational equation by graphing.

6. $\frac{3}{x} + \frac{1}{2} = 1$ _____

7. $\frac{1}{x - 1} + \frac{1}{2} = \frac{5}{2}$ _____

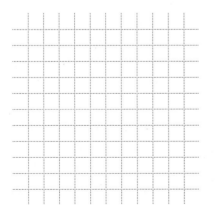

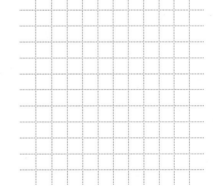

_____

Quick Warm-Up: Assessing Prior Knowledge
11.6 *Proof in Algebra*

Name the property that is illustrated by each statement. Assume that all variables represent real numbers.

1. $3(10 + 7) = 3(10) + 3(7)$ _____

2. $(k + 7)(0) = 0$ _____

3. If $a + 6 = 5$, then $a + 6 + (-6) = 5 + (-6)$. _____

4. If $r = 6 + 8$ and $6 + 8 = 14$, then $r = 14$. _____

Lesson Quiz
11.6 *Proof in Algebra*

1. Identify the hypothesis in the statement "If $\frac{a}{b} = \frac{c}{d}$, then $\frac{a-b}{b} = \frac{c-d}{d}$."

_____

2. Write the converse of the statement "If $a - 5 < 3$, then $a < 8$."

_____

For Exercises 3 and 4, write a proof for each conjecture. Give a reason for each step. Let all variables represent real numbers.

3. For all values of m, $5m + 3m = 8m$. _____   4. If $4\left(x - \frac{1}{2}\right) = 22$, then $x = 6$. _____

State a reason for each step in the proof of the following conjecture:

$$\text{"If } \frac{a}{b} = \frac{c}{d}, \text{ then } \frac{a-b}{b} = \frac{c-d}{d}.\text{"}$$

Statement	Reason
5. $\frac{a}{b} = \frac{c}{d}$	_____
6. $\frac{a}{b} - 1 = \frac{c}{d} - 1$	_____
7. $\frac{a}{b} - \frac{b}{b} = \frac{c}{d} - \frac{d}{d}$	_____
8. $\frac{a-b}{b} = \frac{c-d}{d}$	_____

Chapter Assessment

Chapter 11, Form A, page 1

Write the letter that best answers the question or completes the statement.

_____ 1. Which rational expression is undefined for $x = -1$?

     a. $\dfrac{x+1}{x^2-3}$ b. $\dfrac{x+3}{x^2+1}$ c. $\dfrac{x+3}{x+1}$ d. $\dfrac{x+1}{x-1}$

_____ 2. Which function is graphed at right?

     a. $f(x) = \dfrac{1}{x+1} + 1$

     b. $f(x) = \dfrac{1}{x-1} + 1$

     c. $f(x) = \dfrac{1}{x+1} - 1$

     d. $f(x) = \dfrac{1}{x-1} - 1$

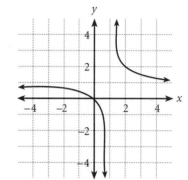

_____ 3. The value of ABS (-6.5) − ABS (6) is

     a. -0.5 b. -12.5 c. 12.5 d. 0.5

_____ 4. Given that y varies inversely as x, if y is $\frac{1}{2}$ when x is 20, what is the value of x when y is 5?

     a. 2   b. $\dfrac{1}{4}$ c. 4   d. $2\frac{1}{2}$

_____ 5. Time varies inversely as the average speed over a given distance. When Aaron travels 3 hours, his average speed is 75 km/h. What is Aaron's speed when his time is 5 hours?

     a. 15 km/h   b. 25 km/h   c. 45 km/h   d. 225 km/h

_____ 6. When simplified, $\dfrac{15a^2b^3}{20ab^5}$ is equal to

     a. $\dfrac{3a}{4b^2}$.   b. $\dfrac{a}{4b}$.   c. $4ab$.   d. $\dfrac{a^3b^8}{4}$.

_____ 7. When simplified, $\dfrac{3x^2-6x}{x^2-3x+2}$ is equal to

     a. $\dfrac{3}{x-1}$.   b. $\dfrac{x-2}{x-1}$.   c. $\dfrac{3x}{x-1}$.   d. 3.

_____ 8. What is the sum $\dfrac{5n}{3} + \dfrac{3}{2n}$?

     a. $\dfrac{5n+3}{6n}$.   b. $\dfrac{5n+3}{5n}$.   c. $\dfrac{10n^2+9}{6n}$.   d. $\dfrac{10n^2+9}{6n^2}$.

_____ 9. The difference $y - \dfrac{y-2}{y+2}$ is equal to

     a. $\dfrac{y^2-y-2}{y+2}$.   b. $\dfrac{y^2+y+2}{y+2}$.   c. $y-1$.   d. $y+1$.

Chapter Assessment

Chapter 11, Form A, page 2

_____ 10. The product $\dfrac{7a^3b}{2c} \cdot \dfrac{c^3}{21a^2b^3}$ is equal to

    a. $\dfrac{2a^2c^2}{3b^2}$.   b. $\dfrac{ac^2}{6b^2}$.   c. $\dfrac{a^3b^4c^4}{42}$.   d. $\dfrac{a^3c^3}{b^3}$.

_____ 11. What is the product of $\dfrac{x^2 - 3x + 2}{x^2 - 1}$ and $\dfrac{x^2 + x}{3x}$?

    a. $\dfrac{x-2}{3}$ b. $\dfrac{x-2}{3x}$ c. $\dfrac{x(x-2)}{3}$ d. $\dfrac{x-2}{3(x+1)}$

_____ 12. The quotient $\dfrac{p^3}{2qr^2} \div \dfrac{-p^2r}{4q}$ is

    a. $-\dfrac{p^5}{8p^2r}$.   b. $-\dfrac{2p}{r}$.   c. $-\dfrac{2}{pr^3}$.   d. $-\dfrac{2p}{r^3}$.

_____ 13. The quotient $\dfrac{w^2 - 6w + 9}{w^2 + 5w + 6} \div \dfrac{4w - 12}{w + 2}$ is

    a. $\dfrac{w-3}{4w-12}$ b. $\dfrac{w-3}{w+3}$ c. $\dfrac{w-3}{4w+12}$ d. $\dfrac{w+3}{w-3}$

_____ 14. For which values of k is $\dfrac{3k}{2k+1} + \dfrac{4}{k} = 0$ true?

    a. $k = \dfrac{2}{3}, k = 2$ b. $k = -\dfrac{2}{3}, k = -2$

    c. $k = \dfrac{3}{4}, k = -4$ d. $k = -\dfrac{3}{4}, k = 4$

_____ 15. The solution of the equation $\dfrac{2}{x-1} - \dfrac{1}{2} = -\dfrac{5}{2}$ is

    a. 2.   b. 0.   c. -2.   d. 1.

_____ 16. The difference between two numbers is 21. If the larger number is divided by the smaller, the quotient is 2 with a remainder of 6. What are the two numbers?

    a. 21 and 42   b. 3 and 19   c. 15 and 36   d. 17 and 38

_____ 17. If $\dfrac{m}{5} = \dfrac{m+2}{3}$, then

    a. $m = 5$.   b. $m = \dfrac{5}{4}$.   c. $m = -1$.   d. $m = -5$.

_____ 18. The value of INT(-4.8) is

    a. -5.   b. 5.   c. -4.   d. 4.

_____ 19. Using the definition of an odd number, which of the following shows that $4k + 17$ is an odd number?

    a. $4k + 17 = 2(2k + 8)$ b. $4k + 17 = 2(2k) + 17$

    c. $4k + 17 = 4k + 16 + 1$ d. $4k + 17 = 2(2k + 8) + 1$

Chapter Assessment

Chapter 11, Form B, page 1

1. What condition causes a rational expression to be undefined? _____

2. Explain how the expression $x - 3$ is related to $3 - x$.

   _____

3. In the statement "If $a + 3 > 8$, then $a > 5$," what part is assumed to

   be true? _____

Graph each rational function on the grid provided.

4. $y = \dfrac{1}{x} + 2$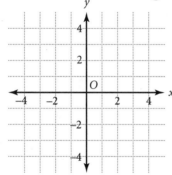

5. $y = -\dfrac{1}{x + 1}$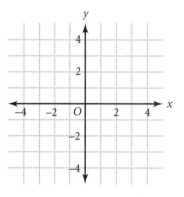

For Exercises 6–8, y varies inversely as x.

6. If y is 8 when x is 3, find x when y is 4. _____

7. If x is 12 when y is 5, find y when x is 15. _____

8. If y is 16 when x is $\dfrac{1}{2}$, find x when y is 1. _____

9. The frequency of a vibrating string is inversely proportional to its
   length. A violin string that is 10 inches long vibrates at a frequency
   of 512 cycles per second. Find the frequency of an 8-inch string.

   _____

Simplify each expression.

10. $\dfrac{14x^4y^2}{21x^3y^3}$ _____

11. $\dfrac{2a^3 - 6a^2}{2a^2 - 18}$ _____

12. $\dfrac{(m - 1)^2}{(1 - m)^2}$ _____

13. $\dfrac{3k + 12}{16 - k^2}$ _____

14. $\dfrac{5n - 10}{n^2 + 3n - 10}$ _____

15. $\dfrac{b^2 - 2b - 15}{b^2 - 6b + 5}$ _____

Algebra 1

Chapter Assessment

Chapter Assessment
Chapter 11, Form B, page 2

Perform the indicated operations.

16. $\dfrac{5}{2x} + \dfrac{7}{6x^2}$ _____

17. $\dfrac{1}{5a} - \dfrac{4}{7a^2}$ _____

18. $\dfrac{3m^5}{4n^2} \cdot \dfrac{12n}{15m^2}$ _____

19. $\dfrac{18p^3}{5q^2} \div \dfrac{9p}{q^3}$ _____

20. $\dfrac{y^2}{2y + 3} + 3$ _____

21. $\dfrac{b^2 - 4b}{b - 1} - \dfrac{3}{1 - b}$ _____

22. $\dfrac{2k^2 - 7k + 3}{2k^2 - 4k} \cdot \dfrac{k^2 - k - 2}{6k - 3}$ _____

23. $\dfrac{(z - 3)^2}{z + 1} \div \dfrac{z^2 - 9}{3z^2 + 3z}$ _____

Solve algebraically.

24. $\dfrac{5}{x} + \dfrac{1}{2} = \dfrac{3}{x}$ _____

25. $\dfrac{x + 2}{x} + \dfrac{2}{3x} = 5$ _____

26. $\dfrac{x + 1}{x^2 - 4} - \dfrac{7}{x + 2} = -3$ _____

27. $\dfrac{5}{x - 3} + 1 = \dfrac{4}{x^2 - 4x + 3}$ _____

28. Use the grid provided to solve $-\dfrac{2}{x} + \dfrac{1}{2} = 3$ by graphing.

_____

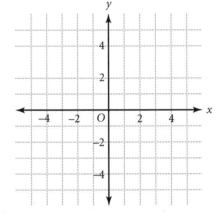

Evaluate.

29. ABS (-5.8) _____

30. INT$\left(3\dfrac{2}{3}\right)$ _____

31. Jill and her younger sister Samantha mow lawns to earn extra money. If Jill can mow a lawn in 45 minutes and Samantha can mow a lawn in $1\dfrac{1}{2}$ hours, how long will it take if they work together?

_____

32. Write the hypothesis for the converse of the statement "If the length of the side of a square is $x + 3$, then the perimeter is $4x + 12$."

_____

Alternative Assessment

Operations With Rational Expressions, Chapter 11, Form A

TASK: Dscribe the operations of addition, subtraction, multiplication, and division with rational expressions.

HOW YOU WILL BE SCORED: As you work through the task, your teacher will be looking for the following:

- how well you understand the concept of operations with rational expressions
- whether you can describe the processes of addition, subtraction, multiplication, and division with rational expressions
- how well you can perform the indicated operations

Refer to the fractions $\frac{2b^2 - b}{6b}$ and $\frac{4b^2 + b}{3b}$.

1. Explain why these two fractions are rational expressions.

_____

_____

2. For what value(s) of the variable are the fractions undefined? Explain.

_____

_____

3. How can you find a common denominator for these two rational expressions? Find the sum and difference of these rational expressions.

_____

_____

4. Describe how to find the product of these rational expressions. Find the product, and explain whether it is in simplest form.

_____

_____

5. Explain the process used to divide these rational expressions. Divide the first rational expression by the second. Then describe how you can check your result.

_____

SELF-ASSESSMENT: Choose two rational expressions and perform the four arithmetic operations on them. Explain any difficulties you had in solving any of the problems.

Alternative Assessment

Formulating Problems By Using Graphs, Chapter 11, Form B

TASK: Formulate your own time, work, investment, or real-world problem.

HOW YOU WILL BE SCORED: As you work through the task, your teacher will be looking for the following:

- how well you can work backward to formulate your own problems
- whether you can write the equations that represent the solution shown by a graph

1. Examine the first graph. Write a time problem described by the rational equation $\frac{14}{x} = \frac{9}{x-5}$.

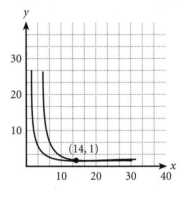

2. Examine the second graph. Write a work problem described by the rational equation $9 = \frac{60 + 4x}{x}$.

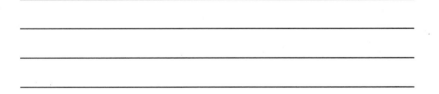

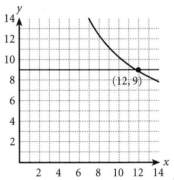

3. Describe the third graph. Then write an investment problem represented by the equation $t = \frac{72}{r}$. (Hint: This is the graph of the function for the Rule of 72.)

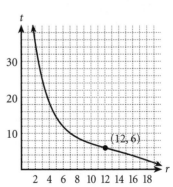

SELF-ASSESSMENT: Create your own graph to represent a real-world problem.

Quick Warm-Up: Assessing Prior Knowledge
12.1 Operations With Radicals

Evaluate.

1. 14^2 _____

2. $(-5)^2$ _____

3. $\sqrt{81}$ _____

4. $-\sqrt{1}$ _____

Simplify.

5. $y \cdot y$ _____

6. $p(p-2)$ _____

7. $(k+3)(k-3)$ _____

8. $(n-4)(n-1)$ _____

Lesson Quiz
12.1 Operations With Radicals

Find each square root. If the square root is irrational, estimate the value to the nearest hundredth.

1. $\sqrt{169}$ _____

2. $\sqrt{50}$ _____

3. $\sqrt{2500}$ _____

4. $\sqrt{48}$ _____

5. $\sqrt{72}$ _____

6. $\sqrt{45}$ _____

7. $\sqrt{49}$ _____

8. $\sqrt{54}$ _____

9. $\sqrt{75}$ _____

Express in simplest radical form.

10. $\sqrt{5} \cdot \sqrt{20}$ _____

11. $\sqrt{6} \cdot \sqrt{12}$ _____

12. $\sqrt{3} \cdot \sqrt{32}$ _____

13. $\dfrac{\sqrt{9}}{\sqrt{25}}$ _____

14. $\dfrac{\sqrt{48}}{\sqrt{3}}$ _____

15. $\dfrac{\sqrt{72}}{\sqrt{8}}$ _____

Simplify each of the following. Assume that all variables are nonnegative and that all denominators are nonzero.

16. $\sqrt{a^2b^4}$ _____

17. $\sqrt{\dfrac{m^5}{n^4}}$ _____

18. $\sqrt{x^5y^7}$ _____

19. $\sqrt{s^3t^6}$ _____

20. $\sqrt{bc^8}$ _____

21. $\sqrt{\dfrac{g^9}{h^5}}$ _____

If possible, perform the indicated operations and simplify your answer.

22. $2\sqrt{3} + 5\sqrt{3}$ _____

23. $8\sqrt{2} - 3\sqrt{2}$ _____

24. $\sqrt{48} + \sqrt{75}$ _____

25. $\sqrt{45} - \sqrt{20}$ _____

26. $\dfrac{4 + \sqrt{20}}{2}$ _____

27. $\dfrac{\sqrt{12} + \sqrt{48}}{\sqrt{3}}$ _____

Quick Warm-Up: Assessing Prior Knowledge

12.2 Square-Root Functions and Radical Equations

Solve each equation.

1. $2m + 9 = 6$ _____

2. $s^2 = 36$ _____

3. $(a + 2)^2 = 9$ _____

4. $g^2 + 4g = 0$ _____

5. $u^2 + 4u = 5$ _____

Lesson Quiz

12.2 Square-Root Functions and Radical Equations

Solve each equation algebraically. Be sure to check your solutions.

1. $\sqrt{x + 3} = 5$ _____

2. $\sqrt{5 - x} = 8$ _____

3. $\sqrt{2x + 15} = 5$ _____

4. $\sqrt{4x - 7} = 9$ _____

5. $\sqrt{3x - 2} = x$ _____

6. $\sqrt{12 - x} = x$ _____

7. $\sqrt{x^2 + 5x - 10} = x$ _____

8. $\sqrt{x + 11} = x - 1$ _____

Graph each side of the following equations on the grid provided.
Solve the equations by finding the points of intersection.

9. $\sqrt{x - 1} = 3$

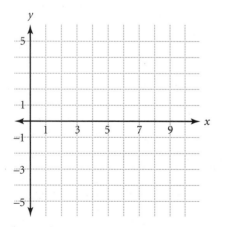

10. $\sqrt{x + 3} = 5$

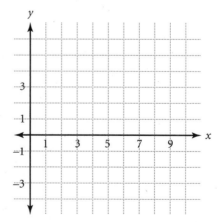

_____   _____

Quick Warm-Up: Assessing Prior Knowledge
12.3 The Pythagorean Theorem

Solve each equation for *x*.

1. $x^2 = 49$ _____

2. $x^2 = 48$ _____

3. $x^2 + 144 = 169$ _____

4. $x^2 + 3^2 = 5^2$ _____

5. $x^2 = y^2 + z^2$ _____

6. $x^2 + y^2 = z^2$ _____

Lesson Quiz
12.3 The Pythagorean Theorem

Complete the table. Use a calculator and round your answers to the nearest tenth.

	Leg	Leg	Hypotenuse
1.	12	16	
3.		12	13
5.	9		18
7.	7	9	
9.		30	34

	Leg	Leg	Hypotenuse
2.	15		17
4.	0.6	0.8	
6.		10	11
8.	11		61
10.	5		13

Solve for *x*. Round answers to the nearest tenth, if necessary.

11.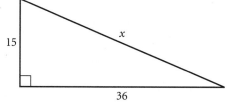

_____

12.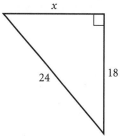

_____

13. A hiker leaves camp and walks 8 miles due west and 12 miles due north. About how far is the hiker from camp? _____

Mid-Chapter Assessment
Chapter 12 (Lessons 12.1–12.3)

Write the letter that best answers the question or completes the statement.

_____ 1. The square root of $\sqrt{0.04}$ is

      a. 0.4.   b. 0.02.   c. 0.002.   d. 0.2.

_____ 2. Which function is graphed below?

      a. $y = -\sqrt{x} + 3$

      b. $y = \sqrt{x-1} + 3$

      c. $y = -\sqrt{x-1} + 3$

      d. $y = -\sqrt{x+3} - 1$

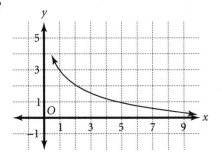

_____ 3. Which of the following statements is false?

      a. $\sqrt{a}\sqrt{b} = \sqrt{ab}$ b. $\sqrt{a+b} = \sqrt{a} + \sqrt{b}$

      c. $\sqrt{a} + \sqrt{b} = \sqrt{b} + \sqrt{a}$ d. $\sqrt{\dfrac{a}{b}} = \dfrac{\sqrt{a}}{\sqrt{b}}$

_____ 4. When simplified $\sqrt{x^4 y^5}$ is equal to

      a. $x^2 y^2\sqrt{y}$.   b. $x^2 y\sqrt{y^3}$.   c. $x^2 y^3$.   d. $xyx^3 y^4$.

Simplify each of the following:

5. $\sqrt{225}$ _____   6. $\sqrt{98}$ _____   7. $\sqrt{90{,}000}$ _____

8. $\sqrt{6}\sqrt{8}$ _____   9. $\dfrac{\sqrt{75}}{\sqrt{3}}$ _____   10. $\dfrac{6}{\sqrt{2}}$ _____

Perform the indicated operation, and simplify your answer.

11. $5\sqrt{3} + 3\sqrt{3}$ _____   12. $\sqrt{24} + \sqrt{54}$ _____

13. $\dfrac{15 + \sqrt{75}}{5}$ _____   14. $\dfrac{\sqrt{12} + \sqrt{15}}{\sqrt{3}}$ _____

Solve for x.

15. _____

16. _____

Quick Warm-Up: Assessing Prior Knowledge

12.4 The Distance Formula

In a right triangle, *a* and *b* are the lengths of the legs and *c* is the length of the hypotenuse. Find each unknown length. Give exact answers.

1. $a = 6, b = 8, c =$ _____

2. $a = \sqrt{2}, b = \sqrt{7}, c =$ _____

3. $a = 1, b = 7, c =$ _____

4. $a = 15, c = 17, b =$ _____

5. $b = \sqrt{7}, c = 4, a =$ _____

6. $a = 1, c = 5, b =$ _____

Lesson Quiz

12.4 The Distance Formula

Find the distance between each pair of points. Give exact answers.

1. $A(5, 9), B(1, 6)$ _____

2. $P(7, 3), Q(0, 8)$ _____

3. $R(-1, 4), S(2, -7)$ _____

4. $C(6, -3), D(-2, 3)$ _____

5. $X(7, -8), Y(9, -8)$ _____

6. $M(-5, 0), N(0, 3)$ _____

Identify the coordinates of the midpoint of each line segment.

7. $P(8, 1), Q(2, 5)$ _____

8. $A(-3, 9), B(1, 3)$ _____

9. $X(5, 0), Y(-1, -4)$ _____

10. $R(6, -2), S(2, -2)$ _____

The midpoint of $\overline{AB}$ is *M*. Calculate the missing coordinates.

11. $A(7, 3), B(\_\_\_\_, \_\_\_\_), M(6, 4)$

12. $A(\_\_\_\_, \_\_\_\_), B(-5, 1), M(-3, 2)$

13. $A(\_\_\_\_, 3), B(2, \_\_\_\_), M(0, -1)$

14. $A(8, \_\_\_\_), B(\_\_\_\_, 1), M(4, -4)$

15. The vertices of $\triangle ABC$ are $A(-3, 5)$, $B(1, 2)$, and $C(0, 9)$. Decide if $\triangle ABC$ is scalene, isosceles, equilateral, or right. Explain your reasoning.

_____

Quick Warm-Up: Assessing Prior Knowledge
12.5 *Geometric Properties*

Find the distance between each pair of points.

1. $R(2, 4), S(2, -7)$ _____

2. $M(-3, -6), N(-1, -6)$ _____

3. $O(0, 0), P(-4, 3)$ _____

4. $A(3, 5), B(-6, 2)$ _____

Find the coordinates of the midpoint of the segment with the given endpoints.

5. $J(1, 7), K(3, 11),$ _____

6. $C(-8, -2), D(2, 8)$ _____

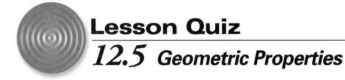

Lesson Quiz
12.5 *Geometric Properties*

Write the equation of a circle with its center at the origin and the given radius.

1. radius of 4 _____   2. radius of 9 _____

3. radius of 5 _____   4. radius of r _____

From each equation of a circle, give the center and the radius.

5. $(x - 3)^2 + (y - 1)^2 = 25$ _____   6. $(x + 5)^2 + y^2 = 36$ _____

7. $(x + 1)^2 + (y + 3)^2 = 16$ _____   8. $x^2 + (y - 2)^2 = 5$ _____

Quick Warm-Up: Assessing Prior Knowledge
12.6 *The Tangent Function*

Write as a decimal.

1. $\dfrac{5}{8}$ _____

2. $\dfrac{5.2}{13}$ _____

Solve each equation.

3. $20 = \dfrac{w}{4}$ _____

4. $20 = \dfrac{4}{a}$ _____

Given $\triangle ABC \sim \triangle XYZ$, $AB = 6$, $AC = 12$, $BC = 9$, and $XY = 3$, find each length.

5. XZ _____

6. YZ _____

Lesson Quiz
12.6 *The Tangent Function*

Use a calculator to find the tangent of each angle to the nearest hundredth.

1. $75°$ _____

2. $7°$ _____

3. $30°$ _____

4. $14°$ _____

5. $37°$ _____

6. $89°$ _____

Find the measure of the angle whose tangent is given. Give each angle measure to the nearest whole number of degrees.

7. 1 _____

8. 1.73 _____

9. 0.21 _____

10. 0.957 _____

11. 1.5 _____

12. 4.25 _____

Find the tangent of each acute angle in the triangles below.

13.

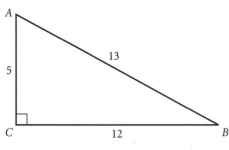

_____

14.

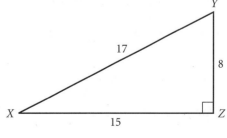

_____

Quick Warm-Up: Assessing Prior Knowledge
12.7 *The Sine and Cosine Functions*

Use a calculator to evaluate the following:

1. tan 57° to the nearest ten-thousandth _____

2. $\tan^{-1}(0.8)$ to the nearest degree _____

In △ABC, m∠C = 90°, AC = 24, and BC = 10. Find the following:

3. tan A _____   4. tan B _____

5. m∠A _____   6. m∠B _____

Lesson Quiz
12.7 *The Sine and Cosine Functions*

Find each of the following for the given triangle:

1. cos D _____   2. sin E _____

3. tan E _____   4. sin D _____

For Exercise 5–8, use right triangle *GHI* at right. Find the indicated length or angle measure to the nearest tenth.

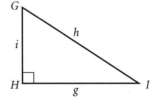

5. $g = 10, h = 15$, m∠I = _____

6. $g = 8, h = 11$, m∠G = _____

7. m∠G = 46°, $h = 18, i =$ _____

8. m∠I = 39°, $h = 5, i =$ _____

Find each of the following.

9. $\cos^{-1}0.2589$ _____   10. $\sin^{-1}1$ _____

11. sin 85° _____   12. cos 55° _____

Algebra 1

Quick Warm-Up: Assessing Prior Knowledge
12.8 Introduction to Matrices

Evaluate.

1. $-15 + 6$ _____

2. $5 + (-3.7)$ _____

3. $4 - 6$ _____

4. $-3 - (-2.1)$ _____

5. $(-12)(4)$ _____

6. $(-9)(-7)$ _____

7. $2(-5) + (-1)(0)$ _____

8. $(-3)(-8) + (-9)(2)$ _____

Lesson Quiz
12.8 Introduction to Matrices

Find the value of each variable in matrices *A* and *B*.

Let $A = \begin{bmatrix} 2w & 19 & -8 \\ 5y-2 & 1 & -4 \end{bmatrix}$ and $B = \begin{bmatrix} 10 & 3x+1 & -8 \\ 43 & 1 & 0.5z \end{bmatrix}$. Matrices A and B are equal.

1. $w =$ _____ 2. $x =$ _____ 3. $y =$ _____ 4. $z =$ _____

Perform each matrix operation on matrices *L*, *M*, and *N*.

Let $L = \begin{bmatrix} -5 & 10 \\ 2 & -8 \end{bmatrix}$, $M = \begin{bmatrix} 1.8 & 4 \\ 0 & 3.2 \end{bmatrix}$, and $N = \begin{bmatrix} \frac{2}{3} & -9 \\ 6 & \frac{1}{2} \end{bmatrix}$.

5. $L + M$　　　　　6. $L - M$　　　　　7. $L + N$　　　　　8. $N - L$

_____　　_____　　　　　　　　_____

9. The Acme Clothing store uses the stock and the sales matrices shown at right for the month of May. Write a new matrix that reflects the stock remaining at the end of May.

_____

	Stock (May 1st)		Sales (May)	
jeans	jackets	125　　82		112　　56
shirts	ties	93　　128		83　　15
pants	socks	46　　234		28　　142

Chapter Assessment

Chapter 12, Form A, page 1

Write the letter that best answers the question or completes the statement.

_____ 1. What is $\sqrt{40}$ to the nearest tenth?

      a. 1600   b. 20   c. 6.3   d. 2.5

_____ 2. Which function is graphed below?

      a. $y = \sqrt{x-1} + 2$
      b. $y = \sqrt{x+2} - 1$
      c. $y = \sqrt{x-2} - 1$
      d. $y = \sqrt{x+1} - 2$

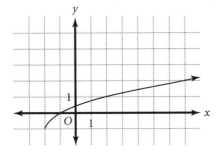

_____ 3. Which of the following statements is false?

      a. $\sqrt{a}\,\sqrt{b} = \sqrt{ba}$ b. $\sqrt{a-b} = \sqrt{a} - \sqrt{b}$ c. $\left(\sqrt{a}\right)^2 - a$ d. $\dfrac{\sqrt{a}}{\sqrt{b}} = \sqrt{\dfrac{a}{b}}$

_____ 4. When simplified, $\sqrt{\dfrac{p^6 q^3}{r^4}} =$

      a. $\dfrac{p^3 q\sqrt{q}}{r^2}.$ b. $\dfrac{p^2 q\sqrt{p^2 a}}{r^2}.$ c. $\dfrac{pq\sqrt{p^4 q}}{r^2}.$ d. $\dfrac{pa\sqrt{q}}{r^2}.$

For Exercises 5–7, perform the indicated operation and write the letter that gives the answer in simplest form.

_____ 5. $\sqrt{24} + \sqrt{54}$

      a. 30   b. $2\sqrt{6} + 3\sqrt{6}$ c. $5\sqrt{6}$ d. $13\sqrt{6}$

_____ 6. $(4\sqrt{18})^2$

      a. 6   b. 36   c. 72   d. 288

Chapter Assessment

Chapter 12, Form A, page 2

_____ 7. $(\sqrt{3} + 8)(\sqrt{3} - 2)$

      a. $11\sqrt{3} - 16$ b. $6\sqrt{3} - 13$ c. $6\sqrt{3} - 16$ d. $8\sqrt{3} - 13$

_____ 8. What is the value of x in $\sqrt{x - 3} = 5$?

      a. 4   b. $3 + \sqrt{5}$ c. 22   d. 28

_____ 9. If $\sqrt{x + 12} = x$, then $x =$

      a. -3 and 4.   b. $\pm\sqrt{12}$.   c. 4.   d. 144.

_____ 10. What is the solution to the equation $4x^2 = 36$?

      a. $x = \pm3$ b. $x = \pm6$ c. $x = 9$ d. $x = 144$

_____ 11. What is the value of x in the given triangle?

      a. 7   b. 17

      c. 23   d. 120

_____ 12. What are the coordinates of the midpoint of segment AB with endpoints $A(6, -1)$ and $B(-2, 1)$?

      a. $(2, 0)$ b. $(4, 1)$ c. $(4, 0)$ d. $(4, 2)$

_____ 13. The midpoint of $\overline{PQ}$ is $M(2, -3)$. If P is the point $(5, -10)$, what are the coordinates of point Q?

      a. $(1, -3)$ b. $(-1, 4)$ c. $(2, -5)$ d. $(1, 4)$

_____ 14. Which of the following is the equation of the circle graphed at right?

      a. $(x - 1)^2 + (y + 2)^2 = 2$

      b. $(x + 1)^2 + (y - 2)^2 = 2$

      c. $(x + 1)^2 + (y - 2)^2 = 4$

      d. $(x - 1)^2 + (y + 2)^2 = 4$

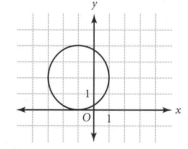

_____ 15. What are the center and radius of the circle $(x + 3)^2 + (y - 5)^2 = 49$?

      a. $(3, -5), r = 49$ b. $(-3, 5), r = 49$ c. $(-3, 5), r = 7$ d. $(3, -5), r = 7$

_____ 16. What are the dimensions of the matrix?

      a. 5×4 b. 4×5 c. 3×2 d. 10×5

_____ 17. What number is in row 2 and column 3 of the matrix?

      a. 8   b. 4   c. 10   d. 6

$$\begin{bmatrix} 10 & 8 & 5 & 9 & 6 \\ 12 & 7 & 6 & 2 & 0 \\ 5 & 8 & 10 & 0 & 1 \\ 9 & 2 & 4 & 3 & 7 \end{bmatrix}$$

Chapter Assessment

Chapter 12, Form B, page 1

1. For what values of x is $\sqrt{-x + 5}$ defined? _____

Find each square root. If the square root is irrational, find the value to the nearest hundredth.

2. $\sqrt{81}$ _____   3. $-\sqrt{225}$ _____   4. $\sqrt{\dfrac{4}{16}}$ _____   5. $\sqrt{32}$ _____

6. Find the length of the side of a square whose area is 49 square centimeters. _____

Graph each function on the grid provided. Give the domain and range.

7. $y = 2\sqrt{x} - 1$

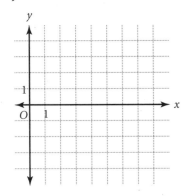

8. $y = \sqrt{x - 3}$

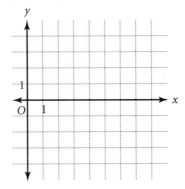

_____   _____

Simplify each of the following:

9. $\sqrt{72}$ _____   10. $\sqrt{0.09}$ _____   11. $\sqrt{243}$ _____

12. $\sqrt{2}\,\sqrt{3}$ _____   13. $(3\sqrt{7})^2$ _____   14. $\sqrt{75}$ _____

15. $\dfrac{\sqrt{28}}{\sqrt{3}}$ _____   16. $5\sqrt{3} - 7\sqrt{3}$ _____   17. $\sqrt{48} + \sqrt{45} - \sqrt{75}$ _____

Solve each equation algebraically.

18. $\sqrt{x - 3} = 7$ _____   19. $\sqrt{3x} = 9$ _____   20. $\sqrt{2x + 5} = 3$ _____

Use matrix *A* for Exercises 21 and 22.

21. Give the entry at A_{24}. _____

22. Give the dimensions of matrix A. _____

$$A = \begin{bmatrix} 12 & 7 & -1 & -2 \\ 5 & 0 & 3 & 9 \end{bmatrix}$$

Chapter Assessment
Chapter 12, Form B, page 2

Solve for x.

23.

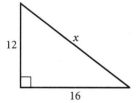

24.

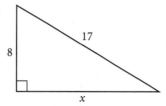

Find the distance between each pair of points.

25. $A(-2, 3), B(6, -3)$ _____   **26.** $P(0, -2), Q(3, 4)$ _____

27. Given $P(-1, 8), Q(-1, 5), R(3, 5)$, determine whether $\triangle PQR$ is a scalene, isosceles, equilateral, or right triangle. _____

The midpoint of *CD* is *M*. Calculate the missing coordinates.

28. $C(3, 6), D(7, 2), M(\_\_\_\_, \_\_\_\_)$ **29.** $C(-3, 1), D(5, -5), M(\_\_\_\_, \_\_\_\_)$

30. $C(4, 1), D(\_\_\_\_, \_\_\_\_), M(3, 2)$ **31.** $C(\_\_\_\_, \_\_\_\_), D(2, 4), M(5, 1)$

From each equation of a circle, give the center and radius.

32. $x^2 + y^2 = 5$ _____   **33.** $(x - 3)^2 + (y + 1)^2 = 49$ _____

34. Graph $(x - 2)^2 + y^2 = 9$ below.

Use the information in the given triangle *ABC* to find the indicated length or angle measure. Give lengths to the nearest hundredth and degree measures to the nearest whole number.

35.

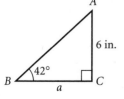

side $a =$ _____

36.

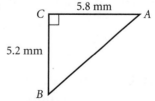

angle $B =$ _____

Alternative Assessment

Properties of Radicals, Chapter 12, Form A

TASK: Explore the properties of adding, subtracting, multiplying, and dividing radicals.

HOW YOU WILL BE SCORED: As you work through the task, your teacher will be looking for the following:

- how well you understand the concept of operations with radicals
- whether you can make generalizations
- whether you can simplify radicals

1. Create two radical expressions, $\sqrt{a}$ and $\sqrt{b}$. Does $\sqrt{a} + \sqrt{b} = \sqrt{a + b}$? Does $\sqrt{a} - \sqrt{b} = \sqrt{a - b}$? Explain why or why not.

_____

_____

2. Use your two radical expressions from Exercise 1. Does $\sqrt{a} \cdot \sqrt{b} = \sqrt{ab}$? Does $\dfrac{\sqrt{a}}{\sqrt{b}} = \sqrt{\dfrac{a}{b}}$? Explain why or why not.

_____

If possible, perform the indicated operations, and simplify your answer.

3. $6\sqrt{3} - 2\sqrt{2}$

4. $\dfrac{\sqrt{12} - \sqrt{6}}{\sqrt{3}}$

5. $5\sqrt{2} + 2\sqrt{5}$

_____   _____   _____

6. Write a generalization about the conditions necessary for adding radical expressions.

_____

7. Describe how factoring is used to simplify $\sqrt{98}$.

_____

8. Write the procedure for multiplying $(2\sqrt{2} + 3)(\sqrt{3} - 2)$. Then find the product.

_____

SELF-ASSESSMENT: Use a calculator to explore fourth roots by finding the square roots of square roots. Repeat Exercises 1 and 2. Write a generalization about what you discover.

Alternative Assessment

Connecting Geometry and Algebra, Chapter 12, Form B

TASK: Solve a real-world problem by using the Pythagorean Theorem.

HOW YOU WILL BE SCORED: As you work through the task, your teacher will be looking for the following:

- whether you understand the problem
- how well you can show the steps used to solve a problem
- whether you can calculate the solution to a problem

Demane lives 33 miles due east of the WNJN radio station. While driving due north from his house, he was able to keep the radio signal for about 56 miles. What is the broadcasting range of WNJN?

1. Draw a diagram to illustrate the problem.

2. Assign a variable to the unknown length in your diagram.

3. Explain how to find the missing length in your diagram.

_____

4. Solve the problem.

_____

5. Describe how you can check your solution.

_____

_____

6. Draw your diagram on graph paper. Use $A(0, 0)$ as the location of the radio station. Explain how to find an equation that models the broadcasting range of WNJN.

_____

_____

SELF-ASSESSMENT: Use coordinate geometry to find the broadcasting range of WNJN. Then compare the different methods to find the solution to the problem. Which did you prefer and why?

Quick Warm-Up: Assessing Prior Knowledge
13.1 *Theoretical Probability*

Write as a percent.

1. $\frac{15}{30}$ _____

2. $\frac{30}{48}$ _____

3. Two number cubes were rolled 120 times. A sum of 7 appeared 18 times. Based on these results, what is the experimental probability that a sum of 7 will appear on the next roll of

   these number cubes? _____

Lesson Quiz
13.1 *Theoretical Probability*

An integer between 1 and 35, inclusive, is drawn at random. Find the probability that the integer is

1. odd. _____

2. even. _____

3. a multiple of 5. _____

The letters in the word *theoretical* are written on index cards and placed in a brown bag. A letter is selected at random.

4. What is the probability of selecting a consonant? _____

5. What is the probability of selecting the letter *t* ? _____

6. What is the probability of selecting a vowel? _____

7. Raymond rolls two number cubes. He needs a sum of 8 to land on the last square and win the game. What is the probability that he

   will win? _____

8. A number cube is rolled once. What is the probability of getting a

   number less than 3? _____

Quick Warm-Up: Assessing Prior Knowledge
13.2   Counting the Elements of Sets

A fair number cube is rolled once. Find the probability of each event below.

1. 1 _____

2. an even number _____

3. 4 _____

4. a prime number _____

A fair number cube is rolled two times. Find the probability of each event below.

5. two 3s _____

6. a sum of 5 _____

Lesson Quiz
13.2   Counting the Elements of Sets

List the integers from 1 to 20 inclusive that are

1. multiples of 3. _____

2. multiples of 6. _____

3. multiples of 3 AND multiples of 6. _____

4. multiples of 3 OR multiples of 6. _____

In a class of 33 students, 10 students have blue eyes, and 17 students have blond hair, including 8 who have both.

5. Show this information in a Venn diagram. (Include all 33 students.)

6. How many students have either blue eyes OR blond hair? _____

7. How many students have neither blue eyes NOR blond hair? _____

8. What is the probability of rolling a 2 OR an even number on one

roll of a number cube? _____

Quick Warm-Up: Assessing Prior Knowledge
13.3   The Fundamental Counting Principle

How many outcomes are in the sample space for each experiment?

1. flipping a coin once _____

2. flipping a coin twice _____

3. flipping a coin three times _____

4. rolling a number cube once _____

5. rolling a number cube twice _____

Lesson Quiz
13.3   The Fundamental Counting Principle

1. The Beagles are ordering new baseball jerseys. There are 7 styles of jerseys from which to choose. For each style, there are 3 different ways the team name and numbers could be affixed to the jersey. How many combinations are possible?

_____

2. Points *A*, *B*, and *C* lie on the same line. How many ways are there to name the line using two of the letters?

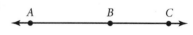

_____

3. Jamal is buying stereo components. He can choose from 3 receivers, 6 CD players, 4 cassette players, and 5 kinds of speakers. How many combinations of components are possible?

_____

Mid-Chapter Assessment

Chapter 13 (Lessons 13.1–13.3)

Write the letter that best answers the question or completes the statement.

_____ 1. Let n be the number of elements in a sample space. Let f be the number of favorable outcomes in the sample space. Which formula would you use to find the theoretical probability, P?

    a. $P = \frac{n}{f}$ **b.** $f = \frac{P}{n}$ **c.** $n = \frac{f}{p}$ **d.** $P = \frac{f}{n}$

_____ 2. Which integers from 1 to 10 inclusive are odd OR multiples of 2?

    a. 1, 2, 3, 4, 5, 6, 7, 8, 9, 10   **b.** 2, 4, 6, 8, 10

    c. 1, 2   **d.** 1, 3, 5, 7, 9

_____ 3. Which is the sample space for tossing two coins?

    a. HH, HH, HH, HH   **b.** HT, TH, HT, HH

    c. HH, HT, TH, TT   **d.** TT, TT, TT, TT

_____ 4. How many ways can you choose an outfit from 4 shirts, 2 pair of jeans, and 3 types of shoes?

    a. 11   **b.** 5   **c.** 12   **d.** 24

5. Eight tickets marked 1 through 8 are placed in a hat. You draw one ticket. What is the probability that the number on the ticket is 4? _____

6. Eight tickets marked from 1 through 8 are placed in a hat. You draw one ticket. What is the probability that the number on the ticket is greater than 4? _____

7. How many elements are in the sample space for a roll of two number cubes? _____

8. In a group of 45 students, 15 take French, 23 take American literature, and 12 take both subjects. Show this information in a Venn diagram.

Quick Warm-Up: Assessing Prior Knowledge
13.4   *Independent Events*

**A card is drawn from a standard deck of 52 playing cards. Find
the probability of each event.**

1. 5 _____

2. heart _____

3. 5 OR heart _____

4. 5 AND hearts _____

**A fair number cube is rolled once. Find the probability of each
event.**

5. 1 _____

6. even _____

7. 1 OR even _____

8. 2 OR even _____

Lesson Quiz
13.4   *Independent Events*

**A bag contains 5 green, 3 orange, and 4 yellow tennis balls. A ball
is drawn and replaced. A second ball is drawn.**

1. What is the probability that both are green? _____

2. What is the probability that both are yellow? _____

3. What is the probability that the first is orange AND the second is green? _____

A red number cube and a green number cube are rolled together.

4. What is the probability that you roll a 3 on the red number cube

   AND a 4 on the green number cube? _____

5. What is the probability that you roll a sum of 7 on the two number cubes? _____

6. What is the probability that you roll a 5 on the red number cube

   OR a 4 on the green number cube? _____

7. What is the probability that you roll a sum of less than 5 on the two

   number cubes? _____

NAME _____ CLASS _____ DATE _____

Quick Warm-Up: Assessing Prior Knowledge
13.5   Simulations

A number cube was rolled 120 times. Based on the results in the table, find the experimental probability of each event below.

Event	Occurrences
1	18
2	21
3	17
4	20
5	23
6	21

1. 4 _____

2. 1 _____

3. even _____

4. odd _____

5. 1 OR 4 _____

6. 4 OR even _____

Lesson Quiz
13.5   Simulations

Nico makes a strike or a spare 60% of the time when bowling. He rolls the ball 20 times during a game. The spreadsheet below shows a simulation of 10 trials. Each row represents one trial. Let 1 mean that Nico makes a strike or a spare, and let 0 mean that he does not.

U11			=INT(RAND( )*2)																	
	A	B	C	D	E	F	G	H	I	J	K	L	M	N	O	P	Q	R	S	T	U
1	Trial	Results																			
2	1	1	0	0	1	0	0	1	1	0	1	1	1	1	0	1	1	1	0	0	0
3	2	0	0	1	1	1	1	0	1	0	0	1	0	1	0	0	1	0	0	0	0
4	3	1	1	1	1	0	1	0	1	0	0	0	1	1	1	1	1	0	0	0	0
5	4	1	0	1	0	0	1	0	1	1	1	1	1	0	0	0	1	0	1	1	1
6	5	0	0	0	1	0	1	0	0	0	0	0	1	0	1	0	0	1	1	0	
7	6	1	1	1	0	0	1	1	1	1	0	0	0	0	1	1	1	0	1	0	1
8	7	0	0	1	0	0	0	1	1	1	0	0	1	0	0	1	0	0	1	0	0
9	8	1	0	1	1	1	1	1	0	0	1	1	0	1	1	0	0	0	1	1	1
10	9	1	1	1	0	0	0	0	1	0	0	0	0	1	0	1	0	0	0	1	0
11	10	1	0	0	1	1	0	0	0	0	0	0	0	1	1	1	1	0	0	1	0

1. How many strikes or spares did Nico make in the third trial? _____

2. In how many trials did Nico make a strike or spare more than 40% of the time? _____

3. What is the experimental probability that Nico makes 5 strikes or spares in a row? _____

Chapter Assessment

Chapter 13, Form A, page 1

Write the letter that best answers the question or completes the statement.

_____ 1. A coin is tossed three times. You are interested in the probability that exactly two tosses will be tails. What is the number of favorable outcomes?

    a. 3   b. 4   c. 2   d. 8

_____ 2. A box contains 12 tickets numbered 1 through 12. One ticket is drawn at random. What is the probability that the number on the ticket is a multiple of 3?

    a. $\frac{1}{4}$ b. $\frac{1}{3}$ c. $\frac{1}{2}$ d. $\frac{1}{12}$

_____ 3. Which simulation is the best method for finding the experimental probability of a 60% chance of snow?

    a. Toss a coin 60 times and record the sequence of heads or or tails.
    b. Put 6 slips of paper marked *snow* and 4 slips of paper marked *no snow* into a bag and draw them at random.
    c. Roll a number cube 100 times and record how many times a 6 is rolled.
    d. Generate a sequence of random integers from 1 to 40.

_____ 4. To find the probability when you know the size of the sample space and the number of favorable outcomes, which of the following steps should you take?

    a. Add the number of favorable outcomes and the size of the sample space.
    b. Multiply the number of favorable outcomes by the size of the sample space.
    c. Divide the number of favorable outcomes by the size of the sample space.
    d. Divide the size of the sample space by the number of favorable outcomes.

_____ 5. The table below shows the results of a simulation in which one number cube was rolled 10 times. What is the experimental probability of getting a 5 on one roll?

Trials	1	2	3	4	5	6	7	8	9	10
Number	4	3	6	6	1	5	2	6	2	5

    a. $\frac{1}{5}$ b. 5   c. 1   d. $\frac{1}{6}$

Chapter Assessment

Chapter 13, Form A, page 2

_____ 6. Which integers from 1 to 20 inclusive are multiples of 4 AND multiples of 8?

a. 4, 8, 16   b. 4, 8, 12, 16, 20
c. 2, 4, 6, 8, 10   d. 8, 16

_____ 7. A plane seats 3 passengers on each side of the aisle. Passengers seated in rows 20 through 25 were asked to board the plane first. Assuming that each seat was booked, how many passengers was this?

a. 18   b. 36   c. 30   d. 45

_____ 8. An athletic store stocks 15 kinds of basketball shoes in 6 sizes. Each kind of shoe comes in 3 ankle heights—low-, mid-, and high-tops. How many pairs of basketball shoes should the manager order if she wants to carry one of each combination?

a. 270   b. 24   c. 90   d. 51

_____ 9. Points X, Y, Z, A, and B lie on the same line. How many ways are there to name the line by using 2 letters?

a. 5   b. 4
c. 20   d. 25

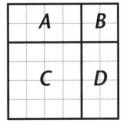

_____ 10. An integer from 1 to 30 inclusive is drawn at random. Find the probability that the number drawn is a multiple of 6.

a. $\frac{1}{5}$ b. 5   c. 6   d. $\frac{1}{6}$

_____ 11. An integer from 1 to 20 inclusive is drawn at random. What is the probability the number drawn is even AND a multiple of 4?

a. 4   b. 5   c. $\frac{1}{4}$ d. $\frac{1}{5}$

_____ 12. What is the probability that when two number cubes are rolled, the sum is less than 8?

a. $\frac{1}{4}$ b. $\frac{7}{36}$ c. $\frac{1}{6}$ d. $\frac{2}{3}$

_____ 13. What is the probability for event B shown in the grid pattern at right?

a. 4   b. $\frac{8}{36}$
c. $\frac{1}{9}$ d. $\frac{1}{3}$

Chapter Assessment

Chapter 13, Form B, page 1

Two number cubes were rolled 10 times with the following results:

Trial	1	2	3	4	5	6	7	8	9	10
Cube 1	6	5	4	1	1	5	3	6	4	6
Cube 2	4	3	4	3	2	5	2	2	3	1

According to the data, find the following experimental probabilities:

1. Both number cubes are alike. _____   2. Both number cubes show a 6. _____

3. At least one number cube is a 3. _____   4. Neither number cube is a 4. _____

5. Design a simulation to find the experimental probability that a student answers all 5 questions correctly on a true-false quiz and explain how you would perform it.

_____

_____

A red number cube and a green number cube are rolled together. Find each probability.

6. The red cube shows a 3 and the green cube shows a 5. _____

7. The red cube shows a prime number OR the green cube shows a 2. _____

8. The sum of the numbers rolled is 1. _____

9. The sum of the numbers rolled is 7 OR 11. _____

10. The sum of the numbers rolled is less than or equal to 10. _____

List the integers from 1 to 10 inclusive that are

11. odd. _____   12. multiples of 4. _____

13. odd AND multiples of 4. _____   14. odd OR multiples of 4. _____

15. Make a Venn diagram that shows where to put the states
Connecticut, Arizona, North Carolina, New Mexico, Utah, New
Jersey, and Illinois in sets of eastern states and western states.
Include all 7 states.

Chapter Assessment
Chapter 13 Form B, page 2

16. After Marlena does her chores, she can choose one sports activity and one arts activity. Make a tree diagram of all possible combinations that she can choose.

Sports	Arts
softball	piano
volleyball	ballet
swimming	guitar
golf	

17. Ramon is buying a new computer. There are 4 color monitors, 7 printers, and 6 keyboards from which to choose. How many combinations are possible? _____

An integer between 1 and 20, inclusive, is drawn at random. Find the probability that it is

18. even. _____   19. odd. _____   20. a prime. _____

21. If the letters of the word *pan* are rearranged, find the probability that an anagram is formed (not including *pan*). _____

Find the probability of each situation.

22. A number cube is rolled twice. What is the probability that both rolls are 3s? _____

23. Gino draws a card from a standard 52-card deck. After replacing the first card, he draws another. What is the probability that the two cards he draws are red? (Half of the cards in the deck are red.) _____

Three pairs of socks are in a laundry basket. One pair is blue, one pair is white, and the third pair is striped. Taylor picks a sock, puts it back in the basket and then picks another sock. Find the probability that

24. both socks are striped. _____

25. the first sock is white and the second is striped. _____

26. neither sock is blue. _____

Alternative Assessment

Experimental Probability, Chapter 13, Form A

TASK: Find an experimental probability by using a simulation.

HOW YOU WILL BE SCORED: As you work through the task, your teacher will be looking for the following:

- how well you can design a simulation
- whether you can find an experimental probability by using a simulation

Vincent has 10 CDs in the glove compartment of his car. Six are by rock artists and 4 are by rap artists.

1. What percent of the CDs are by rock artists? What percent are by rap artists? _____

2. If Vincent picks one CD at random, how many times would you expect him to choose rap? _____

3. How many outcomes are possible each time Vincent randomly selects a CD? _____

4. Describe how you can use experimental probability to predict the number of times that Vincent will pick a rap CD if he selects 3 CDs at random.

   _____

   _____

5. Tell how you would select a random-number generator to simulate Vincent selecting 3 CDs at random.

   _____

   _____

6. Design the simulation and explain how you would perform it.

   _____

   _____

7. Describe how to find the experimental probability that the 3 CDs Vincent randomly selects are all by rap artists.

   _____

SELF-ASSESSMENT: Explain what experimental probability is.

Alternative Assessment
Principles of Counting, Chapter 13, Form B

TASK: Find the theoretical probability.

HOW YOU WILL BE SCORED: As you work through the task, your teacher will be looking for the following:

- how well you can draw and use a Venn diagram
- whether you can find the probability of an event

The table shows the results of a survey of single people who live in one-room apartments.

	College student	Full-time job	Total
Men	8	6	14
Women	7	11	18
Total	15	17	32

1. How many of the men surveyed have full-time jobs? How many

   women have full-time jobs? _____

2. How many of the men OR women surveyed have full-time jobs? _____

3. How many of the people surveyed are women AND are college students? _____

4. How many of the people surveyed are men AND are college students? _____

5. How many of the people surveyed are women AND have full-time jobs? _____

6. If one of the people surveyed is chosen at random, find the probability that the person is:

   a. a woman. _____

   b. a college student. _____

   c. a woman AND a college student. _____

   d. a man OR a college student. _____

SELF-ASSESSMENT: Explain the difference between experimental probability and theoretical probability.

Quick Warm-Up: Assessing Prior Knowledge
14.1 *Graphing Functions and Relations*

Evaluate each expression.

1. $r - 8$, for $r = -3$ _____   2. $|j| - 8$, for $j = -3$ _____   3. $m^2 - 8$, for $m = -3$ _____

4. $(y + 2)^2$, for $y = -10$ _____   5. $6p^2$, for $p = 2$ _____

Lesson Quiz
14.1 *Graphing Functions and Relations*

In Exercises 1–3, tell whether a function is represented. Explain.

1.

Windchill

Wind speed (mph)	5	10	15	20	25
Temperature (at 35°F)	33°F	22°F	16°F	12°F	8°F

_____

2. $\{(3, 21), (3, 0), (3, 1)\}$ _____

3.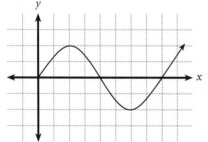

_____

_____

_____

4. Find the domain and range of the function.

$\{(-3, 9), (-2, 4), (-1, 1), (0, 0), (1, 1), (2, 4), (3, 9)\}$ _____

5. Evaluate $g(x) = 2x^2 - 3$ for $x = -1$. _____

Quick Warm-Up: Assessing Prior Knowledge
14.2 Translations

Given $f(x)$ and $g(x)$, find $f(-3)$ and $g(-3)$.

1. $f(x) = -4x$; $g(x) = -4x + 5$ _____

2. $f(x) = x^2$; $g(x) = x^2 + 5$ _____

3. $f(x) = x^2$; $g(x) = (x + 5)^2$ _____

4. $f(x) = |x|$; $g(x) = |x| + 5$ _____

5. $f(x) = |x|$; $g(x) = |x + 5|$ _____

Lesson Quiz
14.2 Translations

1. Describe how the location of the vertex on the graph of
   $y = |x + 3| - 2$ is transformed from the parent function.
   Give the coordinates of the vertex.

   _____

2. Does the function $y = 2x^3 + 1$ represent a horizontal translation or
   a vertical translation? Explain.

   _____

**Identify the parent function and describe any transformations
that are applied to it. Then draw the graph of the function.**

3. $y = -2|x|$

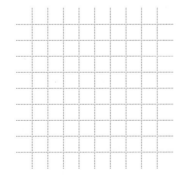

4. $y = -(x - 2)^2$

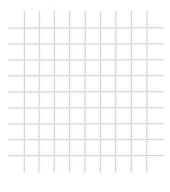

_____   _____

Quick Warm-Up: Assessing Prior Knowledge
14.3 Stretches

Given $f(x)$, find $f(-4)$.

1. $f(x) = x^2$ _____

2. $f(x) = x^2 - 12$ _____

3. $f(x) = 3x^2$ _____

4. $f(x) = 3x^2 - 12$ _____

5. $f(x) = 3(x^2 - 12)$ _____

6. $f(x) = 3(x - 12)^2$ _____

7. $f(x) = 3|x - 12|$ _____

8. $f(x) = 3|x| - 12$ _____

Lesson Quiz
14.3 Stretches

Consider the function $y = \frac{1}{2}x^2$.

1. Identify the parent function. _____

2. Sketch the graph of the function and the parent function on the same grid.

3. How does the graph of $y = \frac{1}{2}x^2$ compare with the graph of the parent function?

_____

_____

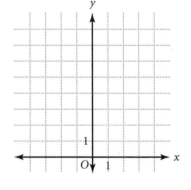

Graph each of the following on the grid provided:

4. $y = \frac{1}{2}|x|$

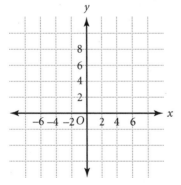

5. $y = 2\left(\frac{1}{x}\right)$

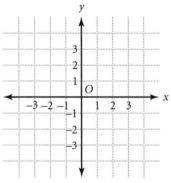

6. How do the y-values of $y = \frac{1}{5}x^2$ compare with the y-values of $y = x^2$?

_____

Mid-Chapter Assessment

Chapter 14 (Lessons 14.1–14.3)

Write the letter that best answers the question or completes the statement.

_____ 1. What is the scale factor of $y = -\frac{1}{3}x + 2$?

     a. 2   b. 3   c. -1 d. $\frac{1}{3}$

_____ 2. Which set of points represents a function?

     a. $\{(1, 2), (3, 5), (1, -2)\}$ b. $\{(1, -1), (1, 0), (1, 1)\}$

     c. $\{(-1, 1), (0, 1), (1, 1)\}$ d. $\{(1, -1), (0, 0), (1, 1)\}$

_____ 3. If $f(x) = 2x^2$, then $f(-1) =$

     a. -3.   b. 3.   c. 2.   d. -1.

_____ 4. Which of the following is an equation for the graph at the right?

     a. $y = 1 - |x|$ b. $y = |x| - 1$

     c. $y = |x - 1|$ d. $y = |x|^{-1}$

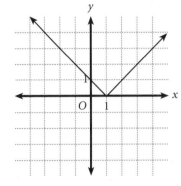

Find the domain and range of each function.

5. $\{(0, 5), (1, 5), (2, 5)\}$ 6. $y = x^2 - 1$

_____   _____

For $f(x) = 2x^2 + 1$, evaluate the following:

7. $f(3)$ _____   8. $f(0)$ _____   9. $f(-2)$ _____   10. $f\left(\frac{1}{2}\right)$ _____

Consider the function $y = \frac{1}{2}x^2$.

11. Identify the parent function. _____

12. Graph the function on the grid provided.

13. Describe the transformation.

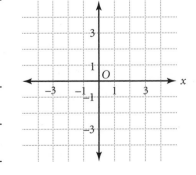

_____

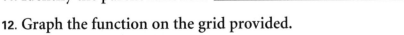

_____

Quick Warm-Up: Assessing Prior Knowledge
14.4 Reflections

Given $f(x)$ and $g(x)$, find $f(6)$ and $g(6)$.

1. $f(x) = 5x; g(x) = -5x$ 2. $f(x) = x^2; g(x) = -x^2$ 3. $f(x) = |x|; g(x) = -|x|$

_____   _____   _____

4. $f(x) = 2^x; g(x) = -2^x$ _____   5. $f(x) = \frac{3}{x}; g(x) = -\frac{3}{x}$ _____

Lesson Quiz
14.4 Reflections

In Exercises 1 and 2, $f(x) = -\frac{1}{x}$.

1. Using x-values of $-2, -1, 0, 1,$ and 2, complete the table of values below.

x	-2	-1	0	1	2
$f(x)$					

2. With the aid of the table from Exercise 1, graph the function $f(x) = -\frac{1}{x}$ on the grid provided.

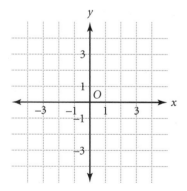

In Exercises 3 and 4, $y = x^2$.

3. Describe the effect of reflecting the graph of $y = x^2$ across the x-axis?

_____

4. Write an equation for the reflection of the graph of $y = x^2$ across the x-axis. _____

Quick Warm-Up: Assessing Prior Knowledge
14.5 Combining Transformations

Identify the transformation of the graph of the parent function
$f(x) = x^2$ **that would result in the graph of each function below.**

1. $g(x) = x^2 - 3$ _____

2. $h(x) = (x + 2)^2$ _____

3. $j(x) = 4x^2$ _____

4. $k(x) = -x^2$ _____

Lesson Quiz
14.5 Combining Transformations

1. Explain the steps that you would carry out in order to sketch the

   graph of the function $y = -2(x - 1)^2$. _____

Sketch the graph of each function on the grid provided.

2. $f(x) = 2|x + 1| - 3$

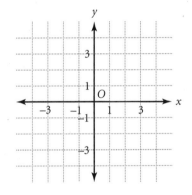

3. $h(x) = -(x - 2)^2 + 1$

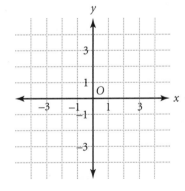

Name the point that corresponds to the point (−2, 4) after
applying the specified transformations to the graph of $y = x^2$.
(Hint: If you are not sure, draw a sketch.)

4. A vertical translation by 2, followed by a vertical reflection. _____

5. A vertical translation by 1, followed by a horizontal translation by 2. _____

6. A vertical stretch by 3, followed by a horizontal translation of −2. _____

7. A vertical reflection, followed by a vertical stretch by $\frac{1}{2}$. _____

Chapter Assessment

Chapter 14, Form A, page 1

Write the letter that best answers the question or completes the statement.

_____ 1. Which of the following represents a function?

a.
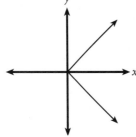

b.
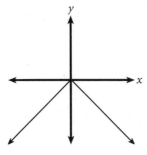

c.
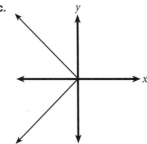

d.
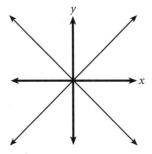

2. What is the range of the function graphed at right?

a. $\{-2, 2, 4\}$ b. $\{0, -4\}$

c. $\{2 \leq x \leq 2\}$ d. $\{-4 \leq y \leq 0\}$

_____ 3. Evaluate $y = -|x + 1| + 3$ for $f(-3)$.

a. 1   b. 5   c. -1 d. 4

_____ 4. Which set of ordered pairs is not a function?

a. $\{(0, 1), (1, 2), (2, 3)\}$ b. $\{(1, 0), (2, 0), (3, 0)\}$

c. $\{(0, 1), (0, 2), (0, 3)\}$ d. $\{(1, 1), (2, 2), (3, 3)\}$

_____ 5. Which is the parent function of the graph at right?

a. $y = |x|$ b. $y = x^2$

c. $y = \dfrac{1}{x}$ d. $y = -\dfrac{1}{x}$

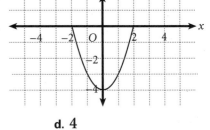

_____ 6. Which kind of transformation of the parent graph is the graph at the right?

a. Reflection   b. Vertical stretch   c. Translation   d. Vertical compression

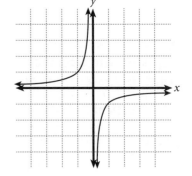

Chapter Assessment

Chapter 14, Form A, page 2

_____ 7. Compared with its parent function, the graph of the function $y = x^2 - 3$ is translated

    a. vertically 3 units down   b. horizontally 3 units to the right
    c. vertically 3 units up   d. horizontally 3 units to the left

_____ 8. Which of the following can be used to identify the amount and direction of the horizontal translation given by $g(x) = a|x - h| + k$?

    a. a b. x c. k d. h

_____ 9. Which function results when the parent function $y = |x|$ is compressed vertically by a factor of $\frac{1}{2}$?

    a. $y = 2|x|$ b. $y = \frac{1}{2}|x|$ c. $y = -2|x|$ d. $y = |x| + \frac{1}{2}$

_____ 10. Which function is a vertical stretch of the parent function?

    a. $y = \sqrt{\frac{1}{3}x}$ b. $y = 3x^2$ c. $y = |x + 3|$ d. $y = (3x)^2$

_____ 11. For the function $y = -2(x + 3)^2 - 1$, the parent function is

    a. $y = x + 3$ b. $y = 2(x + 3)^2$ c. $y = x^2$ d. $y = x^2 - 1$

_____ 12. Which function results when the parent function $y = |x|$ is stretched vertically by a factor of $\frac{1}{2}$, reflected through the x-axis, and translated horizontally 3 units to the left?

    a. $y = -\frac{1}{2}|x| - 3$ b. $y = \frac{1}{2}|-x| + 3$ c. $y = -\frac{1}{2}|x - 3|$ d. $y = -\frac{1}{2}|x + 3|$

_____ 13. What type of transformation is applied to the parent function $y = \frac{1}{x}$ to obtain $y = \frac{1}{x - 3}$?

    a. stretch   b. reflection
    c. horizontal translation   d. vertical translation

_____ 14. The point $(3, 9)$ is on the graph of $y = x^2$. What are the coordinates of the corresponding point on the graph of $y = -x^2 + 3$?

    a. $(3, 12)$ b. $(3, -12)$ c. $(3, 78)$ d. $(3, 84)$

_____ 15. Given $y = -2^x$, what type of transformation is applied to its parent function?

    a. stretch   b. reflection
    c. horizontal translation   d. vertical translation

_____ 16. The price of a jacket was $48. The store put the jacket on sale at half-price and then reduced the sale price by $6. What was the final price of the jacket?

    a. $18   b. $21   c. $27   d. $30

Chapter Assessment

Chapter 14, Form B, page 1

1. What is the difference between a relation and a function?

_____

2. Write $f(-2) = 3$ as an ordered pair so that it can be plotted. _____

Tell whether each of the following represents a function:

3. $(-3, 9), (0, 0), (-3, 5)$

4.

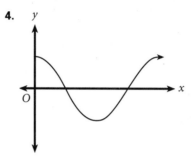

For $g(x) = \dfrac{3}{x-2}$, **evaluate the following:**

5. $g(0)$ _____

6. $g(8)$ _____

7. $g\left(\dfrac{1}{2}\right)$ _____

Evaluate each function for $x = 2$.

8. $h(x) = |x - 1|$ _____

9. $f(x) = x^2 + 2$ _____

10. $g(x) = \dfrac{1}{x} - 2$ _____

11. What happens to the graph of $y = -|x|$ when you reflect it through

the x-axis? _____

Give the parent function for each graph.

12.

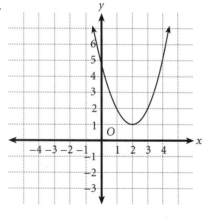

13.

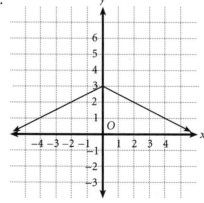

_____   _____

Chapter Assessment

Chapter 14, Form B, page 2

Consider the function $y = \frac{1}{x+2} - 1$.

14. Identify the parent function. _____

15. Sketch the graph of the function.

16. Describe the transformation.

_____

_____

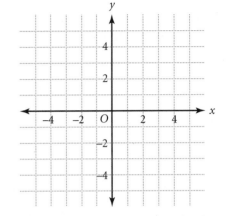

State whether each function is a stretch or compression of the parent function. If so, give the scale factor.

17. $y = 2(x-3)^2 + 1$ _____

18. $y = \frac{1}{2}|x| + 3$ _____

19. $y = \sqrt{3x}$ _____

20. $y = x + 2$ _____

21. Use the grid provided to show what happens to the parabola when you perform a reflection through the y-axis followed by a vertical translation of -3.

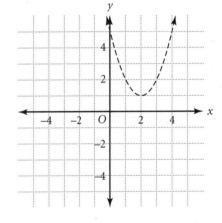

22. Determine the coordinates of the vertex of $y = (x+3)^2 - 1$. _____

Tell whether each function is stretched, compressed, reflected, or translated from its parent function.

23. $y = \frac{x^2}{2}$ _____

24. $y = (x+5)^2$ _____

25. $y = \frac{1}{x-2}$ _____

26. $y = -\frac{1}{x}$ _____

Alternative Assessment

Exploring Transformations, Chapter 14, Form A

TASK: Describe how a change in the function rule and a change in its graph are related.

HOW YOU WILL BE SCORED: As you work through the task, your teacher will be looking for the following:

- whether you can match the graph of ordered pairs from a table to a parent function
- how well you can identify the transformation that is applied to the parent function
- whether you can find the general function that fits the data

Refer to the table of values below.

x	-3	-2	-1	0	1	2	3
y	6	1	-2	-3	-2	1	6

1. Graph the ordered pairs.

2. Describe how to identify the parent function of the graph of the ordered pairs.

   _____

   _____

3. Identify and graph the parent function using the same grid.

   _____

4. Explain how you can compare the graphs of the parent function and the function.

   _____

5. Identify the transformation that was applied to the parent function.

   _____

6. Find a general function that fits the data in the table.

   _____

SELF-ASSESSMENT: Explain how you could have solved the problem by finding the general formula from the number pattern in the data.

Alternative Assessment

Combinations of Transformations, Chapter 14, Form B

TASK: Explore combinations of transformations, stretches, reflections, and translations.

HOW YOU WILL BE SCORED: As you work through the task, your teacher will be looking for the following:

- whether you can explain the steps for sketching a graph
- how well you understand the effect of the order when combining transformations
- how well you can communicate your responses in writing

Sketch the graph of $y = -|x + 3| + 1$.

1. Explain the steps you would perform to sketch this graph.

   _____

   _____

   _____

2. Describe the transformations applied to the parent function of this graph.

   _____

   _____

3. Sketch the graph on the grid provided.

4. Explain whether the order in which you applied the transformations would make a difference in your graph.

   _____

   _____

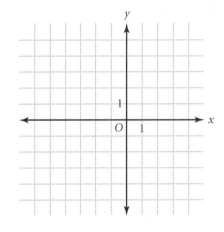

5. Write a generalization about how the order of transformations affects the graph of a function with combined transformations.

   _____

   _____

SELF-ASSESSMENT: Describe how you can use technology to sketch the graph of a function.

Answers

Chapter 1

Quick Warm-Up 1.1

1. 6   2. −5   3. 49   4. 125

Lesson Quiz 1.1

1. 122, 152   2. 95, 80   3. 64, 128

4. 40, 47   5. 14, 12   6. 157, 152

7. first differences: 2, 4, 6; second difference: 2; next 3 terms: 59, 69, 81

8. 182, 192

Quick Warm-Up 1.2

1. 5; 8   2. 20; 24

Lesson Quiz 1.2

1.
b	1	2	3	4	5
4b	4	8	12	16	20

2. 34   3. 9   4. −2   5. $\frac{1}{3}$ 6. 6 tickets

7. $1.80   8. $2.98x = 15$

Quick Warm-Up 1.3

1. 168.3   2. 33.5   3. 27   4. 480

Lesson Quiz 1.3

1. 31   2. 7.35   3. 46   4. $\frac{5}{17}$ 5. 3   6. 10

7. 9   8. 7   9. 42   10. 512   11. 4000

12. 75

13. Add the numbers in parentheses and then divide; 5.

14. Add the numbers in parentheses, then perform the multiplication, and then subtract; 18.

15. Perform the division in parentheses. Then perform the multiplication and then the addition; 11.

Mid-Chapter Assessment—Chapter 1

1. c   2. b   3. d   4. a

5. Multiply each the preceding number by 3.

6. June   7. 20, 23, 26, 29

8.
y	1	2	3	4	5
6y	6	12	18	24	30

9. 1   10. $\frac{1}{4}$

Quick Warm-Up 1.4

1–3.

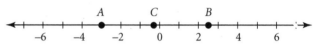

Lesson Quiz 1.4

1–3.

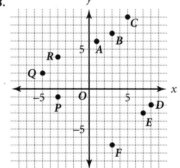

4–7.

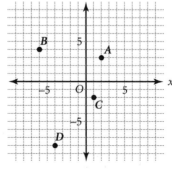

Answers

1. no 2. yes 3. no 4. $(2, 3)$ 5. $(-6, 4)$
6. $(1, -2)$ 7. $(-4, -8)$

8.

x	1	2	3	4	5
y	1	4	7	10	13

Quick Warm-Up 1.5

1. 41, 47, 53 2. 36, 28, 20
3. 125, 253, 509 4. $-1, 3, 7, 11, 15$
5. 7, 13, 19, 25, 31

Lesson Quiz 1.5

1. $y = 5x - 3$ 2. $y = 5 + 2x$

3.
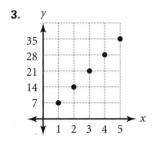

4.
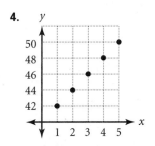

Quick Warm-Up 1.6

1–3.
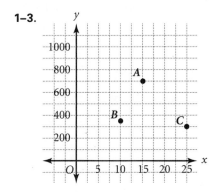

Lesson Quiz 1.6

1. no correlation

2. negative correlation

3. positive correlation

4. Line 1; the points appear to be closer to line 1 than to line 2.

Chapter Assessment, Form A—Chapter 1

1. d 2. c 3. b 4. d 5. d 6. b 7. d
8. c 9. a 10. a 11. d 12. c 13. d
14. b 15. a 16. c 17. d 18. c 19. d

Form B

1. 38, 29, 20 2. 17, 20, 16 3. 51, 58, 56

4. first differences: 4, 6, 8; second difference: 2; next three terms: 65, 77, 91

5.

z	1	2	3	4	5
7z	7	14	21	28	35

6.

y	1	2	3	4	5
3y	3	6	9	12	15

7. $w = 4$ 8. $b = 6$ 9. $x = 8$ 10. $a = 4$
11. 36 12. $28.33\overline{3}$ 13. 8.68 14. 62.48
15. 74 16. 52 17. 45 18. yes 19. yes

20.

x	1	2	3	4	5
y	6	7	8	9	10

21.

x	1	2	3	4	5
y	4	8	12	16	20

22. little to none 23. positive 24. negative

25. Line A; line B rises in the wrong direction.

Answers

Alternative Assessment, Form A—Chapter 1

1. Stage 3: $48, \frac{1}{9}, \frac{16}{3}$

    Stage 4: $192, \frac{1}{27}, \frac{64}{9}$

    Stage 5: $768, \frac{1}{81}, \frac{256}{27}$

2. Sequence 1: 3, 12, 48, 192, 768, 3072, 12,288

3. Sequence 2: $1, \frac{1}{3}, \frac{1}{9}, \frac{1}{27}, \frac{1}{81}, \frac{1}{243}, \frac{1}{729}$

4. Sequence 3:

    $3, 4, \frac{16}{3}, \frac{64}{9}, \frac{256}{27}, \frac{1024}{81}, \frac{4096}{243}$

Since each side of the equilateral triangle produces 4 new sides, you can find the number of sides by multiplying by 4. Since each new snowflake produces sides with lengths $\frac{1}{3}$ of a unit, you can find the new length by multiplying by $\frac{1}{3}$.

To find the perimeter of the snowflake, multiply the number of sides by the length of each side after each stage.

Score Point 4: Distinguished

The student demonstrates a comprehensive understanding of the concept of mathematical sequences and is able to continue the snowflake sequence without computational errors. The student uses perceptive, creative, and complex mathematical reasoning as well as precise and appropriate mathematical language throughout the task. Theoretical knowledge is apparent and is applied to concrete situations as the student successfully demonstrates an ability to visualize and sketch snowflakes at various stages.

Score Point 3: Proficient

The student demonstrates a broad understanding of the concept of mathematical sequences and is able to continue the snowflake sequence with few serious flaws and only minor computational errors. The student uses precise and appropriate language most of the time. Theoretical knowledge is apparent and is applied to concrete situations as the student attempts to visualize and sketch snowflakes at various stages.

Score Point 2: Apprentice

The student demonstrates an understanding of the concept of mathematical sequences and is able to continue the snowflake sequence given specific parameters. The student uses appropriate mathematical language some of the time and applies theoretical knowledge to the task but may not always apply the correct strategies in all situations.

Score Point 1: Novice

The student demonstrates a basic understanding of the concept of mathematical sequences but is unable to complete the task. The student uses little mathematical reasoning or appropriate mathematical language. Theoretical knowledge may appear weak, and many responses may be illogical because directions were followed incorrectly.

Score Point 0: Unsatisfactory

The student fails to make an attempt to complete the task, and the responses only restate the problem.

Answers

Form B

Answers will vary depending on the responses of each individual survey. Research indicates that only women ages 40–60 make their beds on regular basis.

Score Point 4: Distinguished

The student demonstrates a comprehensive understanding of the concepts of gathering data and making scatter plots and is able to accomplish the task. The student uses perceptive, creative, and complex mathematical reasoning and is able to use sophisticated, precise, and appropriate language throughout the task. Theoretical knowledge is apparent and is applied to concrete situations as the student successfully demonstrates an ability to make predictions and draw conclusions from data.

Score Point 3: Proficient

The student demonstrates a broad understanding of the concepts of gathering data and making scatter plots and is able to accomplish the task with few serious flaws. The student uses precise and appropriate language most of the time. Theoretical knowledge is apparent and is applied to concrete situations as the student attempts to make predictions and draw conclusions from data.

Score Point 2: Apprentice

The student demonstrates an understanding of the concepts of gathering data and making scatter plots. The student uses appropriate mathematical language some of the time and applies theoretical knowledge to the task but may not always apply the correct strategies in all situations.

Score Point 1: Novice

The student demonstrates a basic understanding of the concepts of gathering data and making scatter plots but is unable to complete the task. The student uses little mathematical reasoning or appropriate mathematical language. Theoretical knowledge may appear weak, and many responses may be illogical because directions were followed incorrectly.

Score Point 0: Unsatisfactory

The student fails to make an attempt to complete the task, and the responses only restate the problem.

Chapter 2

Quick Warm-Up 2.1

1. 158　　2. 19　　3. 107　　4. 25　　5. 42

Lesson Quiz 2.1

1. $<$　　2. $>$　　3. $<$　　4. $<$　　5. $>$　　6. $<$
7. -3　　8. 7　　9. -8　　10. 0　　11. 6.1
12. 6.8　　13. 5　　14. 9　　15. 0　　16. 1.5
17. 17　　18. $\frac{5}{6}$　　19. -13　　20. 32　　21. 1

Quick Warm-Up 2.2

1. 943　　2. $\frac{17}{14}$, or $1\frac{3}{14}$　　3. 14.1　　4. 318
5. $\frac{8}{15}$　　6. 6.76

Lesson Quiz 2.2

1. 1　　2. -5　　3. -11.6　　4. 6　　5. -3.3
6. -1.5　　7. -1　　8. $-\frac{3}{4}$　　9. $\frac{11}{12}$　　10. 2

Answers

11. -19 12. 3.6 13. 0.7 14. $\frac{1}{12}$ 15. 0

16. 1 17. 7 18. 0 19. 10 20. -4 21. -2

Quick Warm-Up 2.3

1. 9 2. -3.7 3. 0 4. $\frac{1}{3}$ 5. 3

6. -19 7. $-\frac{1}{5}$ 8. 2.62

Lesson Quiz 2.3

1. -3 2. -6 3. 1 4. -2.8 5. 0

6. -12 7. 2 8. 1.8 9. -2.41 10. 0

11. 6 12. 0 13. -6

Quick Warm-Up 2.4

1. 15.58 2. $\frac{1}{10}$ 3. 17 4. $\frac{2}{5}$

Lesson Quiz 2.4

1. $\frac{3}{2}$ 2. $\frac{4}{5}$ 3. $\frac{7}{3}$ 4. $\frac{4}{-2}$ 5. $\frac{-3}{12}$ 6. $\frac{1}{16}$

7. $\frac{-1}{13}$ 8. $\frac{2}{22}$ 9. $\frac{36}{4}$ 10. $\frac{-11}{7}$ 11. $\frac{3}{-18}$

12. $\frac{-4}{1}$ 13. 5 14. -20 15. -4 16. -16

17. -7 18. 12 19. -64 20. 6 21. -1

22. -10 23. $\frac{4}{5}$ 24. 5

Mid-Chapter Assessment—Chapter 2

1. d 2. a 3. c 4. c 5. b 6. c 7. d

8. $\frac{5}{6}$ 9. $\frac{20}{-2}$ 10. $\frac{3}{1}$ 11. $\frac{7}{2}$ 12. $\frac{3}{6}$ 13. $\frac{22}{11}$

Quick Warm-Up 2.5

1. 112 2. 18 3. -3 4. 150 5. 38

Lesson Quiz 2.5

1. 4; Comm. Prop. of Mult.
   25; Assoc. Prop. of Mult.
   $100 \cdot 12$
   1200

2. 89 3. 1600 4. 290 5. 1116

6. Commutative Property of Addition

7. Associative Property of Addition

8. Distributive Property

9. Distributive Property

Quick Warm-Up 2.5

1. 4 2. -11 3. -4

4. -5 5. -15 6. 75

Lesson Quiz 2.6

1. $(6 + 3)w = 9w$

2. $(2 - 1)t = t$

3. $(-5 - 2)r = -7r$

4. $(3 + 6)p = 9p$

5. $\left(\frac{1}{2} + 1\right)k = 1\frac{1}{2}k$

6. $(0.5 - 0.25)s = 0.25s$

7. $(5 - 2)q = 3q$

8. $(27 - 12)g = 15g$

9. $\left(\frac{2}{5} - \frac{4}{5}\right)k = \frac{-2}{5}k$

10. $6x + 8$

11. $5b - 4$

12. $10z - 8$

13. $6.3x + 1.6y$

Answers

14. $9x - 4$

15. $5d + e$

16. $2m + 14n$

17. $4p - 5q + 6r$

Quick Warm-Up 2.7

1. -90 2. 63 3. -2 4. 3

Lesson Quiz 2.7

1. $21x$ 2. $-21x$ 3. $21x^2$ 4. $-7x$

5. $24x$ 6. -6 7. $1 - 2x$ 8. $-3x + 2$

9. $1 + 2x$ 10. a and b   11. a and b

12. 168

Chapter Assessment—Form A—Chapter 2

1. b   2. c   3. a   4. c   5. b   6. d   7. d

8. c   9. b   10. c   11. c   12. d   13. d

14. d   15. b   16. c   17. b   18. a

Form B

1. -8 2. -3 3. -7 4. 57 5. -21

6. 76 7. -34 8. 3200 9. 512 10. 6

11. 127 12. 29 13. $8\frac{1}{3}$ 14. $12y$ 15. $-12y$

16. $4y$ 17. $1 - 2.2x$ 18. $-10x + 0.55$

19. $48x^2$ 20. $14{,}777$ feet

21. Commutative Property of Addition

22. 35 23. $7c + 4$ 24. $6x - 5$ 25. 12

26. 10 27. 45 28. $-8y - 3y; -11y$

29. $-3m + 4m; m$ 30. $b + c$ 31. $8x$

32. $6d - 2$ 33. $11r + 6s$ 34. $v - 5w$

35. Answers may vary. Sample answer: At noon, 19 students were in the cafeteria. By 12:15 P.M., there were 45 students in the cafeteria. How many students entered the cafeteria during that time?

36. Let C represent the number of cartons and let P represent the number of pots. Then $6C + 2P + 3C + 4P = 9C + 6P$.

37. $\$49$ 38. $\$33$ 39. $-20x - 8$

40. $22x$ 41. $3x^2 - 5$

Alternative Assessment—Form A—Chapter 2

1. From 0 go to -4 and then move 5 units to the right.

2. Check students' diagrams; -3.

3. Add their absolute values and write the common sign.

4. Find the difference of their absolute values and write the sign of the number with the greater absolute value.

5. Answers will vary. Possible answer:
$$-4 - 3 = -4 + (-3)$$
$$= -7$$

6. Answers will vary. Possible answer:
$$5 - 2 = 3; 2 - 5 = -3$$

Score Point 4: Distinguished

The student demonstrates a comprehensive understanding of adding and subtracting real numbers. The student uses perceptive, creative, and complex mathematical reasoning as well as precise and appropriate mathematical language throughout the task. Theoretical knowledge is apparent and is applied to concrete situations as the student successfully draws conclusions based on the investigations.

Answers

Score Point 3: Proficient

The student demonstrates a broad understanding of adding and subtracting real numbers. The student uses precise and appropriate language most of the time. Theoretical knowledge is apparent as the student attempts to draw conclusions based on the investigations.

Score Point 2: Apprentice

The student demonstrates an understanding of adding and subtracting real numbers. The student uses mathematical reasoning most of the time, uses appropriate mathematical language some of the time, and attempts to apply theoretical knowledge to the task, but may not be able to draw conclusions based on the investigations.

Score Point 1: Novice

The student demonstrates some understanding of adding and subtracting real numbers. The student uses little mathematical reasoning or appropriate mathematical language. Theoretical knowledge may appear weak, and many responses may be illogical. Many directions are followed incorrectly.

Score Point 0: Unsatisfactory

The student fails to make an attempt to complete the task, and the responses only restate the problem.

Form B

1. 135°F

2. Justin: $84 - (-51) = 135$
   Laura: $88 - (-52) = 140$
   Difference: $140 - 135 = 5$

3. 126°F   4. 11°F   5. January 18; 6°F

Score Point 4: Distinguished

The student demonstrates a comprehensive understanding of adding and subtracting integers. The student uses perceptive, creative, and complex mathematical reasoning as well as precise and appropriate mathematical language throughout the task. Theoretical knowledge is apparent and is applied to concrete situations as the student successfully draws conclusions based on the investigations.

Score Point 3: Proficient

The student demonstrates a broad understanding of adding and subtracting integers. The student uses precise and appropriate language most of the time. Theoretical knowledge is apparent and is applied to concrete situations as the student attempts to draw conclusions based on the investigations.

Score Point 2: Apprentice

The student demonstrates an understanding of adding and subtracting integers. The student uses mathematical reasoning most of the time, uses appropriate mathematical language some of the time, and attempts to apply theoretical knowledge to the task, but may not be able to draw conclusions based on the investigations.

Score Point 1: Novice

The student demonstrates some understanding of adding and subtracting integers but is unable to complete the task. The student uses little mathematical reasoning or appropriate mathematical language. Theoretical knowledge may appear weak, and many responses may be illogical because directions were followed incorrectly.

Answers

Score Point 0: Unsatisfactory

The student fails to make an attempt to complete the task, and the responses only restate the problem.

Chapter 3

Quick Warm-Up 3.1

1. -6 2. 0 3. 28 4. 0

Lesson Quiz 3.1

1. $y = 8$ 2. $y = 15$ 3. $y = -8$
4. $y = 22$ 5. $x = 13$ 6. $x = 13$
7. $x = 7$ 8. $x = -7$ 9. $b = 22$
10. $c = -11$ 11. $y = \frac{5}{6}$ 12. $y = 1\frac{3}{10}$
13. $49{,}116 + 264 = x$; $49{,}380$ miles
14. $59.95 + t + 0.55 = 65$; $\$4.50$

Quick Warm-Up 3.2

1. -42 2. 96 3. 37
4. -0.5 5. -0.45

Lesson Quiz 3.2

1. $x = 6.625$ 2. $y = 294$ 3. $c = -4$
4. $x = -15$ 5. $b = -1.75$ 6. $y = 6.6\overline{6}$
7. $x = -7.5$ 8. $x = \frac{13}{3}$ 9. $x = 3$
10. $x = 2220$ 11. $y = 5\frac{1}{3}$ 12. $y = -473$
13. $x = 4\frac{5}{6}$ 14. $x = 69$ 15. $a = 3.5$
16. $x = 54$ 17. $\$0.56$ 18. 24

Quick Warm-Up 3.3

1. $x = 8$ 2. $x = 26$ 3. $x = -15$ 4. $x = 6$
5. $x = 5$ 6. $x = 2.5$ 7. $x = 12$ 8. $x = 45$

Lesson Quiz 3.3

1. $p = 5$ 2. $x = 8$ 3. $b = -5$ 4. $x = -3$
5. $x = 10$ 6. $y = 2$ 7. $r = -2$
8. $m = -1$ 9. $m = 1.6$ 10. $n = 4.1$
11. $y = 64$ 12. $y = 81$ 13. $z = -40$
14. $n = 6$ 15. 4 16. 6

Mid-Chapter Assessment—Chapter 3

1. b   2. a   3. c   4. b   5. d   6. d
7. $\$50{,}000$ 8. 6 gallons

9. Let p represent profit in dollars and let x represent the number of people attending. Then $p = 8x - (0.5x + 55)$.

10. 74

Quick Warm-Up 3.4

1. $x = 4$ 2. $x = 2$ 3. $x = -3$ 4. $x = 4$
5. $x = 4.11$ 6. $x = 1.83$

Lesson Quiz 3.4

1. $p = 6$ 2. $x = 5$ 3. $c = 9$ 4. $m = 4$
5. $x = 35$ 6. $z = \frac{400}{9}$ 7. $y = 2.5$
8. $b = 4.2$ 9. $x = 12$ 10. $n = 14$
11. $x = 4$ 12. $x = 16$ 13. $c = 8$
14. $y = 2$ 15. 90 16. 30

Quick Warm-Up 3.5

1. $x = 3$ 2. $x = 3$ 3. $x = 1$
4. $x = 6$ 5. $x = 81$

Lesson Quiz 3.5

1. $x = 3$ 2. $p = 9$ 3. $r = 2$ 4. $y = 2$
5. $t = -3$ 6. $x = -1$ 7. $s = 10$
8. $n = 3$ 9. $q = -1$ 10. $x = 5$

Answers

11. $w = -7$ **12.** $c = 1\frac{3}{4}$

13. $2(5x) + 2(x + 10) = 2(3x) + 2(5x - 6)$

14. $x = 8$

Quick Warm-Up 3.6

1. $x = 0$ **2.** $x = 27$ **3.** $x = \frac{7}{6}$

4. true   **5.** false

Lesson Quiz 3.6

1. $-10°C$ **2.** $18.3°C$ **3.** $x = b - a$

4. $c = f + d$ **5.** $l = \frac{A}{w}$ **6.** $C = P - R$

7. $r = \frac{d}{t}$ **8.** $w = \frac{P - 2l}{2}$ **9.** $b = y - 8x$

10. $x = 3y - 3b$ **11.** $s = \frac{3}{2}r - 3$

12. $m = \frac{w + 4n}{8}$ **13.** 4 inches   **14.** $200,000

Chapter Assessment—Form A— Chapter 3

1. a   **2.** b   **3.** a   **4.** b   **5.** d   **6.** d

7. c   **8.** a   **9.** b   **10.** c   **11.** d   **12.** a

13. c   **14.** b   **15.** c   **16.** d   **17.** a   **18.** b

Form B

1. Let x represent the amount of money Tom has. Then $x + 10 = 15$.

2. Let x represent the amount of money Manuel has. Then $x + 5x = 60$.

3. Let c represent the cost of one CD. Then $8c + 1.34 = 105.26$.

4. $b = -12$ **5.** $x = 14$ **6.** $x = 588$

7. $x = 400$ **8.** $c = -0.9$ **9.** $b = 20$

10. $x = 3$ **11.** $y = 20$ **12.** $a = 36.4$

13. $x = -2$ **14.** $x = \frac{1}{3}$ **15.** $x = 4$

16. $m = 3$ **17.** $y = 44$

18. $6x + 25 = 40$, $2.50   **19.** $x = 42.85$

20. $x = 7.6$ **21.** $x = \frac{1}{5}$

22. All real numbers are the solution to this equation.

23. There are no real number solutions to this equation.

24. $5°C$

25. $t = s - r$

26. $c = \frac{ab}{d}$

27. $50°$

28. 13 inches

29. $165

Alternative Assessment—Form A— Chapter 3

1. Read the problem carefully. Identify what is given and what is to be found.

2. Joni's salary is $50 more than Marshall's weekly salary. Thus, Joni's salary is $320 + 50 = 370.

3. Let Billy's salary be represented by b. Then $b = 370 - 20$.

4. $350

5. Decide if the answer is reasonable. Then check the solution. If Billy earns $350, Joni earns $370. Joni's weekly salary of $370 is $50 more than Marshall's weekly salary.

6. Billy's weekly salary is $350.

Answers

Score Point 4: Distinguished

The student demonstrates a comprehensive understanding of addition and subtraction equations. The student uses perceptive, creative, and complex mathematical reasoning as well as precise and appropriate mathematical language throughout the task. Theoretical knowledge is apparent and applied to concrete situations as the student successfully draws conclusions based on the investigations.

Score Point 3: Proficient

The student demonstrates a broad understanding of addition and subtraction equations. The student uses precise and appropriate language most of the time. Theoretical knowledge is apparent and applied to concrete situations as the student attempts to draw conclusions based on the investigations.

Score Point 2: Apprentice

The student demonstrates an understanding of addition and subtraction equations. The student uses mathematical reasoning most of the time as well as appropriate mathematical language some of the time and attempts to apply theoretical knowledge to the task, but may not be able to draw conclusions based on the investigations.

Score Point 1: Novice

The student demonstrates some understanding of addition and subtraction equations but is unable to complete the task. The student uses little mathematical reasoning or appropriate mathematical language. Theoretical knowledge may appear weak and many responses may be illogical because directions were followed incorrectly.

Score Point 0: Unsatisfactory

The student fails to make an attempt to complete the task and the responses only restate the problem.

Form B

1. One plan charges a flat fee and the other charges according to the number of minutes of on-line time.

2. You need to find when the monthly charge would be the same.

3. Let x represent the number of minutes of on-line time.

   $0.05x + 15 = 25$, $x = 200$

4. The companies will cost the same for 200 minutes of on-line time.

5. Answers may vary. A sample is given. I would choose the flat rate so I would not be limited in my on-line time.

Score Point 4: Distinguished

The student demonstrates a comprehensive understanding of the problem and how to model the problem with a multi-step equation. The student uses perceptive, creative, and complex mathematical reasoning as well as precise and appropriate mathematical language throughout the task. Theoretical knowledge is apparent and is applied to concrete situations as the student successfully draws conclusions based on the investigations.

Score Point 3: Proficient

The student demonstrates a broad understanding of the problem and how to model the problem with a multi-step equation. The student uses precise and appropriate language most of the time. Theoretical

Answers

knowledge is apparent and applied to concrete situations as the student attempts to draw conclusions based on the investigations.

Score Point 2: Apprentice

The student demonstrates an understanding of the problem and how to model the problem with a multi-step equation. The student uses mathematical reasoning most of the time as well as appropriate mathematical language some of the time and attempts to apply theoretical knowledge to the task, but may not be able to draw conclusions based on the investigations.

Score Point 1: Novice

The student demonstrates some understanding of the problem and how to model the problem with a multi-step equation but is unable to complete the task. The student uses little mathematical reasoning or appropriate mathematical language. Theoretical knowledge may appear weak and many responses may be illogical as directions are followed incorrectly.

Score Point 0: Unsatisfactory

The student fails to make an attempt to complete the task, and the responses only restate the problem.

Chapter 4

Quick Warm-Up 4.1

1. $\frac{1}{4}$ 2. $\frac{4}{5}$ 3. $\frac{3}{4}$ 4. $\frac{18}{11}$

5. $k = 16$ 6. $n = 18$ 7. $r = 0.125$

8. $y = 30$

Lesson Quiz 4.1

1. $c = 18$ 2. $f = 16$ 3. $h = 108$

4. $t = 7$ 5. $y = 60$ 6. $m = 140$

7. $p = 9.5$ 8. $n = 12$ 9. $z = 1.2$

10. 4   11. 20   12. 17.5 feet

Quick Warm-Up 4.2

1. 0.8   2. 1.375   3. 0.07   4. 1.17

5. $y = 125$ 6. $t = 0.6$ 7. $s = 2.25$

8. $a = 150$

Lesson Quiz 4.2

1. 0.35   2. 0.125   3. 0.006   4. $\frac{3}{4}$ 5. $1\frac{7}{20}$

6. $\frac{1}{250}$ 7. 20   8. $33\frac{1}{3}\%$ 9. 25%   10. 60

11. 200%   12. 10%   13. 8   14. 125%

15. $49   16. 20%

Quick Warm-Up 4.3

1. 0.75; 75%   2. 0.7; 70%   3. 0.23; 23%

4. 0.125; 12.5%

5. $0.\overline{2}$; $22\frac{2}{9}\%$, or about 22.2%

6. $0.41\overline{6}$; $41\frac{2}{3}\%$, or about 41.7%

Lesson Quiz 4.3

1. 50%   2. 10%   3. 60%   4. 40%

5. 42.5%   6. 70%   7. 56%   8. 40%

9. 50%

Mid-Chapter Assessment—Chapter 4

1. c   2. a   3. c   4. c   5. $m = 6$

6. $b = 16$ 7. $t = 0.5$ 8. 20   9. 30

10. 112   11. $33\frac{1}{3}\%$ 12. 200%   13. 15

14. 39.1   15. 25%   16. 60%

Quick Warm-Up 4.4

1. 1, 4, 5, 7, 9, 14, 15, 19

2. 0.25, 0.5, 1.1, 2.5, 7.8, 11.1

Answers

3. 3.4   4. 10.9

Lesson Quiz 4.4

1. mean = 11.2; median 12; mode = 13; range = 10

2. mean = 86; median 82; mode = 82; range = 21

3. mean = 90.4°F; median 90°F; mode = 90°F; range = 7°F

4. The mean and range would change because 70°F would become the lowest temperature.

5. 32

6. mode = 2 hours; mean = 2.5 hours

Quick Warm-Up 4.5

1. 0.68   2. 0.032   3. 0.74   4. about 0.247
5. 8400   6. 3066

Lesson Quiz 4.5

1. Circle graph; the results can be read quickly and easily.

2. about $130 million

3. *Home Alone* and *Batman*

4. $97,200

5. research

Quick Warm-Up 4.6

1. mean: 15; median: 16, modes: 11, 16; range: 11

Lesson Quiz 4.6

1.

Stem	Leaf
13	9 9 9
14	0 3 3 4 7 9 9
15	1 1 3 6 7 8 9 9 9 9
16	1 1 2 2 3 4

2. mean = 152.6; median 154.5; mode = 159

3. 159; the mode is the best answer because it is the value that occurs most often in the data set.

4.

5. between the median and the upper quartile

6. Put the average speeds on the x-axis and the number of times each speed occurs on the y-axis. Then draw the bars.

Chapter Assessment—Form A—Chapter 4

1. b   2. c   3. a   4. b   5. d   6. d   7. b
8. a   9. b   10. b   11. c   12. b   13. a
14. a   15. d   16. c

Form B

1. $p = 10$ 2. $q = 12$ 3. $r = 16$
4. $16.25   5. 0.72   6. 0.075   7. 1.12
8. $\frac{9}{200}$ 9. $\frac{1}{5}$ 10. $1\frac{1}{25}$ 11. 70   12. 160%
13. 20%   14. 92   15. 10.5   16. 3   17. $86
18. 20%   19. 30%   20. 20%   21. 30%
22. 50 wpm   23. 12 hours   24. 45 wpm

25. The more hours of practice, the faster the typing speed.

26. mean = 23.5; median = 20; no mode, range = 27

27. The mean would decrease to 22.7. The median would change to 19.5. The mode would be 18. The range would not be affected.

Answers

28.

Stem	Leaves
1	4 6 6
2	0 1 3 4 6 7 7 7
3	1 2 3 5
4	0 6
5	5

29. mean = 28.5; median = 27; mode = 27; range = 41

30. upper quartile: 33; lower quartile: 21

31.

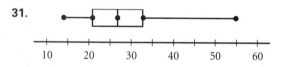

Alternative Assessment—Form A—Chapter 4

1. 4; 8; 12;16

2. The ratio is 3 to 4.

3. Let x be the number of gallons of water. Since the problem represents a proportion, $\frac{3}{4} = \frac{9}{x}$, $3x = 36$, and $x = 12$.

4. Since $3 \cdot 12 = 4 \cdot 9$, the equation is a proportion.

Since all the ratios in the table equal $\frac{3}{4}$, the equation is a proportion.

5. 12   **6.** $2\frac{2}{3}$

Score Point 4: Distinguished

The student demonstrates a comprehensive understanding of using proportions to solve problems. The student uses perceptive, creative, and complex mathematical reasoning as well as precise and appropriate mathematical language throughout the task. Theoretical knowledge is apparent and is applied to concrete situations as the student successfully draws conclusions based on the investigations.

Score Point 3: Proficient

The student demonstrates a broad understanding of using proportions to solve problems. The student uses precise and appropriate language most of the time. Theoretical knowledge is apparent and is applied to concrete situations as the student attempts to draw conclusions based on the investigations.

Score Point 2: Apprentice

The student demonstrates an understanding of using proportions to solve problems. The student uses mathematical reasoning most of the time, uses appropriate mathematical language some of the time, and attempts to apply theoretical knowledge to the task, but may not be able to draw conclusions based on the investigations.

Score Point 1: Novice

The student demonstrates some understanding of using proportions to solve problems but is unable to complete the task. The student uses little mathematical reasoning or appropriate mathematical language. Theoretical knowledge may appear weak, and many responses may be illogical because directions were followed incorrectly.

Score Point 0: Unsatisfactory

The student fails to make an attempt to complete the task, and the responses only restate the problem.

Form B

1. Ice skating: 400 calories per hour; tennis: 420 calories per hour; waterskiing: 480 calories per hour; hill climbing; 490 calories per hour; football: 500 calories per hour; both displays yield the same estimates.

Answers

2. Answers may vary. Sample answer: Graph A emphasizes the similarity among the activities, while Graph B emphasizes the contrast among the activities.

3. The box-and-whisker plot displays the median, the upper and lower quartiles, and the least and greatest values in a set of data.

Score Point 4: Distinguished

The student demonstrates a comprehensive understanding of determining which graph is the best display for a set of data. The student uses perceptive, creative, and complex mathematical reasoning as well as precise and appropriate mathematical language throughout the task. Theoretical knowledge is apparent and is applied to concrete situations as the student successfully draws conclusions based on the investigations.

Score Point 3: Proficient

The student demonstrates a broad understanding of determining which graph is the best display for a set of data. The student uses precise and appropriate language most of the time. Theoretical knowledge is apparent and is applied to concrete situations as the student attempts to draw conclusions based on the investigations.

Score Point 2: Apprentice

The student demonstrates an understanding of determining which graph is the best display for a set of data. The student uses mathematical reasoning most of the time, uses appropriate mathematical language some of the time, and attempts to apply

theoretical knowledge to the task, but may not be able to draw conclusions based on the investigations.

Score Point 1: Novice

The student demonstrates some understanding of determining which graph is the best display for a set of data but is unable to complete the task. The student uses little mathematical reasoning or appropriate mathematical language. Theoretical knowledge may appear weak, and many responses may be illogical because directions were followed incorrectly.

Score Point 0: Unsatisfactory

The student fails to make an attempt to complete the task, and the responses only restate the problem.

Chapter 5

Quick Warm-Up 5.1

1–6.

Lesson Quiz 5.1

1. no   2. yes   3. yes   4. yes

5. $(0, 7)$ 6. $(6, 37)$ 7. $(13, 72)$

8. $(-3, -8)$ 9. $(-10, -43)$ 10. $(3, 22)$

11. $(2, 17)$ 12. $(4, 27)$ 13. $(-2, -3)$

Quick Warm-Up 5.2

1. -5 2. 10 3. -3 4. 0

Answers

Lesson Quiz 5.2

1. 2 2. $-\dfrac{5}{3}$ 3. $\dfrac{2}{3}$ 4. $-\dfrac{3}{4}$ 5. $-\dfrac{7}{2}$ 6. $\dfrac{1}{2}$

7. -1 8. $-\dfrac{4}{5}$ 9. $\dfrac{4}{5}$ 10. 0 11. $-\dfrac{3}{2}$

12. $-\dfrac{1}{9}$

Quick Warm-Up 5.3

1. $t = 0.2$ 2. $m = 70$ 3. $d = 108$

4. $r = 12.6$

Lesson Quiz 5.3

1. 1

2. 0.2 miles per seconds, or 72 miles per hour

3. about 60 miles per hour

4. 5 minutes

5. 20 minutes

Mid-Chapter Assessment

1. d   2. b   3. d   4. c   5. 2 6. $-\dfrac{3}{4}$

7. $-\dfrac{3}{5}$ 8. 0 9. $-\dfrac{1}{4}$ 10. -2.4

11. $\dfrac{2}{5}$; $y = \dfrac{2}{5}x$ 12. 2.8; $y = 2.8x$

13. $\dfrac{1}{2}$; $y = \dfrac{1}{2}x$ 14. $\dfrac{4}{3}$; $y = \dfrac{4}{3}x$

Quick Warm-up 5.4

1. -3 2. 0 3. -1 4. 3 5. -4 6. 1.5

Lesson Quiz 5.4

1. c   2. a   3. b   4. -5; 3 5. $\dfrac{2}{3}$; $-\dfrac{7}{3}$

6. $y = -3x + 5$ 7. $y = -7x + 7$

8. $y = -4x + 3$ 9. $y = -\dfrac{1}{4}x + \dfrac{1}{4}$

10. $y = 2x - 5$ 11. $y = -x - 4$

Quick Warm-Up 5.5

1. $s = t - r$ 2. $g = \dfrac{h}{-5}$

3. $m = \dfrac{p + q}{n}$ 4. $y = \dfrac{z - 4x}{2}$

Lesson Quiz 5.5

1. $3x - y = -2$ 2. $3x - 5y = 10$

3. $5x - 3y = 6$ 4. $2x - y = 8$

5. $y = 3x - 1$ 6. $y = -x + 5$

7. $y = \dfrac{-1}{2}x + 6$ 8. $y = \dfrac{-3}{2}x + \dfrac{9}{2}$

9. x-intercept: 5; y-intercept: -5

10. x-intercept: 3; y-intercept: 2

11. x-intercept: -10; y-intercept: 4

12. x-intercept: 1.2; y-intercept: 3.6

13. $2x + 3y = 7$

14. $y = 1$

15. $5x - y = 13$

16. $x - 3y = -3$

Quick Warm-Up 5.6

1. 1 2. -1 3. slope: -2; y-intercept: 5

4. slope: -1; y-intercept: 2.5

Lesson Quiz 5.6

1. Sample answer: $y = 2x - 3$

2. Sample answer: $y = -\dfrac{1}{2}x$

Answers

1.–2.

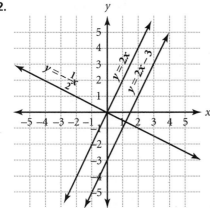

3. $y = \frac{1}{2}x - 2$ **4.** $y = -2x + 3$

5. $y = -x + 3$ **6.** $y = x + 1$

Chapter Assessment—Form A— Chapter 5

1. a   **2.** d   **3.** c   **4.** c   **5.** b   **6.** d   **7.** a

8. c   **9.** c   **10.** b   **11.** a   **12.** c   **13.** b

14. b   **15.** c   **16.** b

Form B

1. 0   **2.** $\frac{1}{8}$ **3.** $y = -x$ **4.** $y = \frac{4}{3}x$

5. $y = 4x$ **6.** $y = \frac{3}{5}x$, or $y = 0.6x$

7. 0   **8.** -2 **9.** $y = 4x - 3$

10. $y = -x - 7$ **11.** $y = -\frac{2}{3}x + 1$

12. $y = -2x$ **13.** $y = \$8.50x$ **14.** \$76.50

15. $x - 2y = 8$ **16.** $y = -2x + 6$

17. 5 miles per hour

18. $y = -\frac{1}{2}x - 1$

19. $y = -3x + 10$

20. $y = -\frac{2}{3}x - \frac{2}{3}$

21. Graphs may vary. The equation should be in the form $y = 3x + b$.

22. Graphs may vary. The equation should be in the form $y = -\frac{4}{3}x + b$.

23.

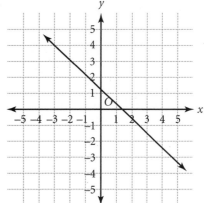

24.

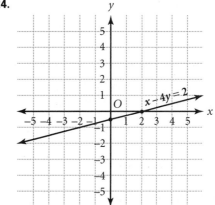

Alternative Assessment—Form A— Chapter 5

1. skier A $= \frac{1}{3}$; skier B $= -\frac{1}{4}$

2. The line for skier A has a positive slope, and the line for skier B has a negative slope.

3. The greater the absolute value of the slope, the steeper the line.

4. skier A: $y = \frac{1}{3}x + 4000$

skier B: $y = -\frac{1}{4}x + 4000$

5. Answers will vary

Answers

Score Point 4: Distinguished

The student demonstrates a comprehensive understanding of positive and negative slope. The student uses perceptive, creative, and complex mathematical reasoning as well as precise and appropriate mathematical language throughout the task. Theoretical knowledge is apparent and applied to concrete situations as the student successfully draws conclusions based on the investigations.

Score Point 3: Proficient

The student demonstrates a broad understanding of positive and negative slope. The student uses precise and appropriate language most of the time. Theoretical knowledge is apparent and applied to concrete situations as the student attempts to draw conclusions based on the investigations.

Score Point 2: Apprentice

The student demonstrates an understanding of positive and negative slope. The student uses mathematical reasoning most of the time as well as appropriate mathematical language some of the time and attempts to apply theoretical knowledge to the task, but may not be able to apply conclusions based on the investigations.

Score Point 1: Novice

The student demonstrates some understanding of positive and negative slope but is unable to complete the task. The student uses little mathematical reasoning or appropriate mathematical language. Theoretical knowledge may appear weak and many responses may be illogical as directions are followed incorrectly.

Score Point 0: Unsatisfactory

The student fails to make an attempt to complete the task, and the responses only restate the problem.

Form B

1. 40 kilometers per hour

2. 40 kilometers per hour

3. The cars are traveling at the same speed.

4. 20

5. Car 1 was 20 kilometers ahead of car 2 at the beginning of the rally.

6. $d - 50 = 40(t - 1); 40t - d = 0$

7.

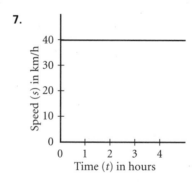

8. It is horizontal; $s = 40$

9. $t = 3$

Score Point 4: Distinguished

The student demonstrates a comprehensive understanding of slope and line relationships. The student uses perceptive, creative, and complex mathematical reasoning as well as precise and appropriate mathematical language throughout the task. Theoretical knowledge is apparent and applied to concrete situations as the student successfully draws conclusions based on the investigations.

Answers

Score Point 3: Proficient

The student demonstrates a broad understanding of slope and line relationships. The student uses precise and appropriate language most of the time. Theoretical knowledge is apparent and applied to concrete situations as the student attempts to draw conclusions based on the investigations.

Score Point 2: Apprentice

The student demonstrates an understanding of slope and line relationships. The student uses mathematical reasoning most of the time as well as appropriate mathematical language some of the time and attempts to apply theoretical knowledge to the task, but may not be able to apply conclusions based on the investigations.

Score Point 1: Novice

The student demonstrates some understanding of slope and line relationships but is unable to complete the task. The student uses little mathematical reasoning or appropriate mathematical language. Theoretical knowledge may appear weak and many responses may be illogical as directions are followed incorrectly.

Score Point 0: Unsatisfactory

The student fails to make an attempt to complete the task and the responses only restate the problem.

Chapter 6

Quick Warm-Up 6.1

1. 0 2. 2 3. -1 4. $\dfrac{2}{3}$ 5. $1\dfrac{1}{3}$ 6. $-1\dfrac{2}{3}$

Lesson Quiz 6.1

1. $x < 14$ 2. $x \le 13$ 3. $x < \dfrac{1}{3}$ 4. $x \ge \dfrac{1}{3}$

5. $t \le 10$

6. $u > 2.5$

7. $v > \dfrac{1}{8}$

8. $m \le -6$

9. $x \ge -2$ 10. $x > 1$ 11. $180 \le x \le 199$

Quick Warm-Up 6.2

1. $t = 3.125$ 2. $b = -2.6$ 3. $a = -0.75$

4. $z = -27$

Lesson Quiz 6.2

1. $x \le 20$ 2. $f \ge -4$ 3. $x > \dfrac{1}{2}$ 4. $h > -4$

5. $x < 9$ 6. $c < -3$ 7. $x \ge 6$

8. $x \ge \dfrac{5}{7}$ 9. $x < 50$ 10. $y > 6$

11. $x < 14$ 12. $y < -2$

13. $17.99x + 15 \le 125$

14. 6

Quick Warm-Up 6.3

1. $h < 9$ 2. $m \ge 5.5$

3. $x \ge 4$

4. $y < -3$

Answers

Lesson Quiz 6.3

1. number line with open circles, marks −1 0 1 2 3 4

2. number line with marks 1 2 3 4 5 6

3. number line with marks 0 2 4 6 8, 7.5

4. number line with marks 2 3 4

5. number line with marks 2 3 4 5 6

6. number line with marks −20 −10 0 10 20, −17, 15

7. number line with marks −23, −4

8. number line with marks −7, 2

9. number line with marks 3.25, 20

10. $-7 \le x < -2$ 11. $p \le -17$ or $p > \dfrac{-1}{5}$

12. $\dfrac{1}{3} < j < 3$ 13. $t < -4$ or $t > -2$

14. $-1\dfrac{1}{2} \le u \le 1$ 15. $-\dfrac{1}{5} < j < 2$

Mid-Chapter Assessment

1. c 2. a 3. d 4. a 5. c
6. b 7. $x \le -2$ 8. $x \le 7$

9. Let x represent the number of concert tickets. $16x \ge 2000$
10. They need to sell at least 125 concert tickets.

Quick Warm-Up 6.4

1. 11 2. 4 3. 4

4. 6 5. 5

Lesson Quiz 6.4

1. 7 2. 12 3. 6 4. 6 5. 20 6. 0

7. Domain: all real numbers Range: $y \ge 0$

8. Domain: all real numbers Range: $y \le 0$

9. Domain: all real numbers Range: $y \ge 3$

10. Domain: all real numbers Range: $y \ge 0$

11.
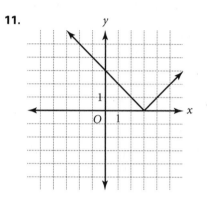

Quick Warm-Up 6.5

1. 6 2. -2 3. -5.1 4. 0

5. $-y - 3$ 6. $-2x + 9$

Lesson Quiz 6.5

1. 1.3 and 1.4 2. 3 and 13 3. -4 and 12

4. $x < -11$ or $x > 5$

5. $-11 \le x \le 1$

6. $x \le -1$ or $x \ge 9$

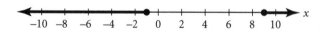

Chapter Assessment—Form A— Chapter 6

1. b 2. d 3. b 4. c 5. d 6. a

Answers

7. a   8. b   9. a   10. c   11. d

12. b   13. d   14. a   15. b   16. c

17. c   18. a

Form B

1. Let x represent the number of people at the party. Then $x > 50$.

2. Let x represent the amount in the savings account. Then $x \leq 450$ and $x \geq 0$; $0 \leq x < 450$.

3. $x > -10$ 4. $y \leq 11$ 5. $c > 1.7$

6. $t > \frac{1}{4}$ 7. $x \leq 50$ 8. $a > 0$ 9. $v \geq 5$

10. False   11. $t \geq -25$ 12. $z < 4$

13. $v \geq 5$ 14. $h > 14$ 15. 6   16. 6

17. $y = 4$ or $y = 2$ 18. $x + 18.50 \leq 30$

19. $3 \leq x \leq 6$

20. $x > 1.5$

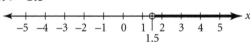

21. $x \leq -5$

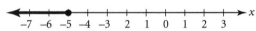

22.

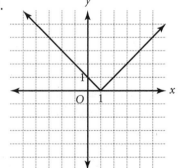

23. Domain: all real numbers
Range: $y \geq 0$

24. $x > 8$ or $x < -4$

25. $5 \leq x \leq 9$

26. $x = 1$ and $x = 6$

27. $x = 14$ and $x = -4$

28. $x \geq -9$

29. $-4 < x < 3$

Alternative Assessment—Form A— Chapter 6

1. 3704 pounds

2. You can write an inequality $x > 3704 + 200$ or $x > 3904$ Blazer, Caprice, Lumina Van

3. $155 \leq$ horsepower ≤ 330

4. $96 \leq w \leq 116$

5. Check the chart to see if all models have at least 155 horsepower. The statement is true.

6. Identify the models that weigh less than or equal to 4000 pounds.

Score Point 4: Distinguished

The student demonstrates a comprehensive understanding of solving problems by solving inequalities. The student uses perceptive, creative, and complex mathematical reasoning as well as precise and appropriate mathematical language throughout the task. Theoretical knowledge is apparent and applied to concrete situations as the student successfully draws conclusions based on the investigations.

Score Point 3: Proficient

The student demonstrates a broad understanding by solving inequalities. The student uses precise and appropriate language most of the time. Theoretical knowledge is apparent as the student attempts to draw conclusions based on the investigations.

Answers

Score Point 2: Apprentice

The student demonstrates an understanding by solving inequalities. The student uses mathematical reasoning most of the time as well as appropriate mathematical language some of the time and attempts to apply theoretical knowledge to the task, but may not be able to apply conclusions based on the investigations.

Score Point 1: Novice

The student demonstrates some understanding of solving inequalities. The student uses little mathematical reasoning or appropriate mathematical language. Theoretical knowledge may appear weak and many responses may be illogical. Many directions are followed incorrectly.

Score Point 0: Unsatisfactory

The student fails to make an attempt to complete the task and the responses only restate the problem.

Form B

1. Between $88 and $108
2. $88 \leq x \leq 108$
3. $|x - 98| \leq 10$
4. Possible answer: I would use the inequality $88 \leq x \leq 108$ and double each part of the inequality.
5. The food budget should be between $100 and $200.
6. Possible answer: The total budget can be found by adding the lodging and food budgets. The inequality $276 \leq x \leq 416$ describes the total budget.

Score Point 4: Distinguished

The student demonstrates a comprehensive understanding of interpreting compound and absolute value inequalities. The student uses perceptive, creative, and complex mathematical reasoning as well as precise and appropriate mathematical language throughout the task. Theoretical knowledge is apparent and applied to concrete situations as the student successfully draws conclusions based on the investigations.

Score Point 3: Proficient

The student demonstrates a broad understanding of interpreting compound and absolute value inequalities. The student uses precise and appropriate language most of the time. Theoretical knowledge is apparent as the student attempts to draw conclusions based on the investigations.

Score Point 2: Apprentice

The student demonstrates an understanding. The student uses mathematical reasoning most of the time as well as appropriate mathematical language some of the time and attempts to apply theoretical knowledge to the task, but may not be able to apply conclusions based on the investigations.

Score Point 1: Novice

The student demonstrates some understanding. The student uses little mathematical reasoning or appropriate mathematical language. Theoretical knowledge may appear weak and many responses may be illogical. Many directions are followed incorrectly.

Score Point 0: Unsatisfactory

The student fails to make an attempt to complete the task and the responses only restate the problem.

Answers

Chapter 7

Quick Warm-Up 7.1

1. $y = 4x - 7$

2. $y = -0.5x + 4$

Lesson Quiz 7.1

1. $(2, 1)$

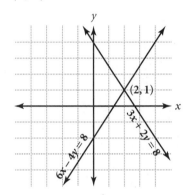

2. $(-3, 2)$

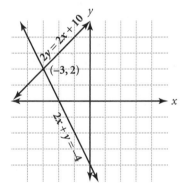

3.

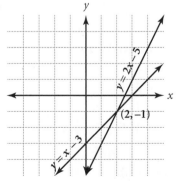

$(2, -1)$ is a solution

Quick Warm-Up 7.2

1. $7r + 15$ 2. $-z + 4$ 3. $18p - 35$
4. $y = -2x + 5$ 5. $y = 3x - 6$

Lesson Quiz 7.2

1. $(-1, 3)$ 2. $(2, 1)$ 3. $(4, 1)$
4. $(-5, -2)$ 5. $(2.2, -1)$ 6. $(1.2, -0.4)$

7. Let x represent the smaller number and y represent the larger number; $x + y = 52$ and $x = y - 12$; 20 and 32

8. $2l + 2w = 120$ and $l = 3w$; The width is 15 meters, and the length is 45 meters.

Quick Warm-Up 7.3

1. $-x$ 2. $-4x - 12y$ 3. $(3, 6)$
4. $(-2, 3)$

Lesson Quiz 7.3

1. Elimination method; there are no terms with positive 1 or negative 1 as a coefficient.

2. Substitution method; the equations are in a form such that an equivalent expression can be substituted for one of the variables.

3. Substitution or elimination method; both y-terms have coefficients of 1, and the equations are in a form such that an equivalent expression can be substituted for one of the variables.

4. $(-2, 3)$ 5. $(5, -1)$ 6. $(3, 3)$
7. $(6, 7)$ 8. $(-1, -4)$ 9. $(2, 5)$

Mid-Chapter Assessment

1. c   2. b   3. c   4. d   5. $(-1, 2)$
6. $(-3, -9)$ 7. $(1, 2)$ 8. $(4, 5)$
9. blouse: $25; skirt: $30

Answers

Quick Warm-Up 7.4

1. $a = 8$

2. all real numbers

3. no real solutions

Lesson Quiz 7.4

1. any multiple of the equation $y = 3x - 5$

2. any equation in the form $y = \frac{1}{2}x + b$, $b \neq 3$

3. consistent

4. inconsistent

5. infinite number of solutions

6. $(3, 2)$ 7. $(8, -1)$

8. no solution

Quick Warm-Up 7.5

1. $s \leq -1$;

2. $m > 2$;

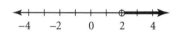

3. $b < -2$ or $b > 3$;

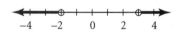

4. $-4 \leq q < 3$;

Lesson Quiz 7.5

1. a solid line with a slope of $-\frac{1}{3}$ and a y-intercept of 2

2. Shade the region on the side of the boundary line that includes $(0, 0)$.

3.

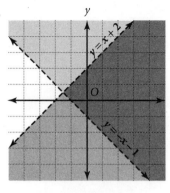

4.

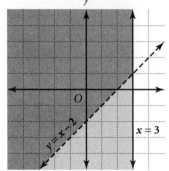

5. $\begin{cases} 2x + 3y \leq 6 \\ 3x + 2y \geq 2 \end{cases}$

Quick Warm-Up 7.6

1. $3r = 150$

2. $8 = p + c$

3. $0.25q = 11.75$

Lesson Quiz 7.6

1. 250 milliliters

2. 15 nickels and 21 quarters

3. 92

4. boat: 4 miles per hour; current: 1 mile per hour

5. Judy is 30 years old, and Phyllis is 10 years old.

Answers

Chapter Assessment—Form A—Chapter 7

1. d 2. a 3. c 4. b 5. a 6. d
7. a 8. c 9. a 10. c 11. b
12. d 13. d 14. a 15. b

Form B

1. Multiply each term of the equation $2x - y = 5$ by the same number.
2. The boundary is the dashed line $x + 3y = 4$.
3. $(-2, 0)$

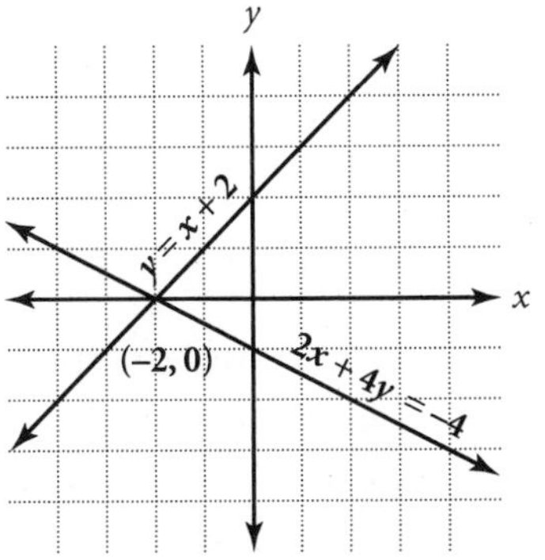

4. infinite number of solutions

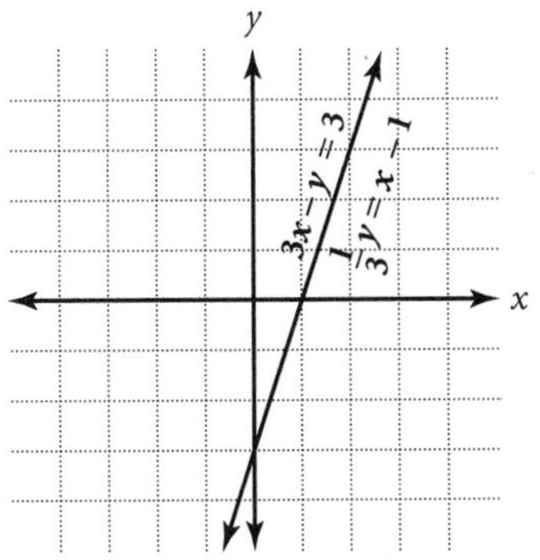

5. $(2, 1)$ 6. $(2, -3)$ 7. $(-1, 2)$
8. $(4.5, -1.5)$ 9. $(1, -1)$
10. no solution
11. 30 and 18
12. width: 10 meters; length: 22 meters
13. consistent and dependent
14. consistent and independent
15. inconsistent
16. $\overleftrightarrow{AB}: x - y = -4$
    $\overleftrightarrow{CD}: 2x + y = -2$
17. any equation with a slope of 1 and a y-intercept not equal to 4
18. any equation whose coefficients are a multiple of those in $2x + y = -2$
19.

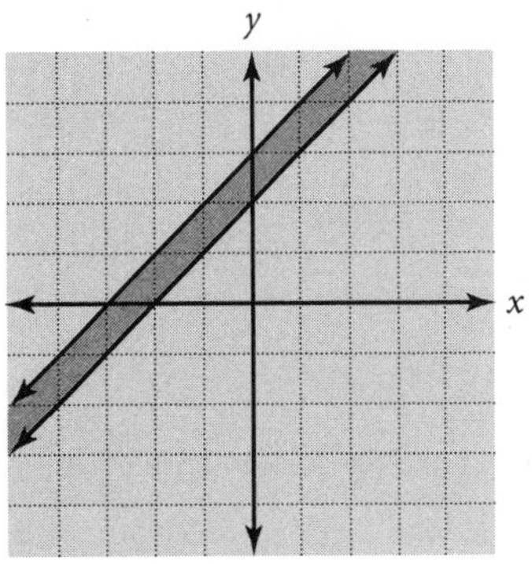

20.

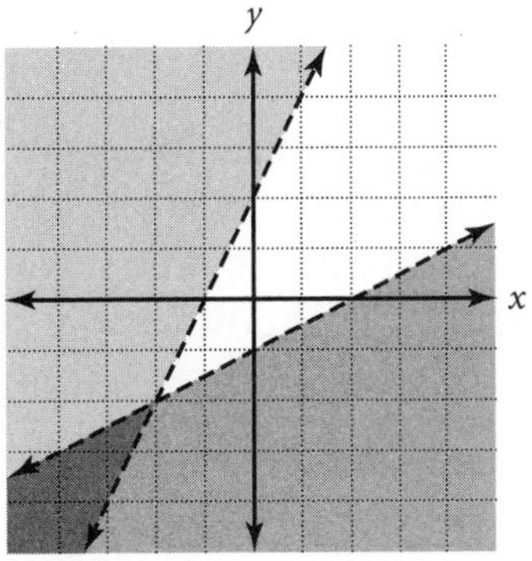

21. 10 dimes and 29 quarters
22. John is 24, and his brother is 16.
23. The speed of the plane is 550 miles per hour, and the rate of wind is 50 miles per hour.

Copyright © by Holt, Rinehart and Winston. All rights reserved.

Answers

Alternative Assessment—Form A—Chapter 7

1.

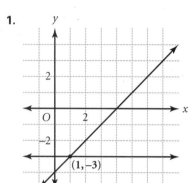

$(1, -3)$

2.

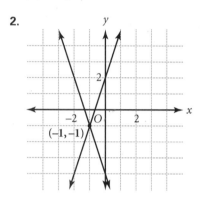

$(-1, -1)$

3. Possible answer: Determine the point of intersection of two lines. Substitute the coordinates of the point of intersection into the equations. This point is the only solution of the system.

4. $(2, -2)$ 5. $(-10, -4)$

6. no solution

7. Solve the second equation for x. Substitute $y + 4$ for x in the first equation. Solve for x, and then substitute x into the other equation and solve for y.

Score Point 4: Distinguished

The student demonstrates a comprehensive understanding of solving equations by graphing and elimination. The student uses perceptive, creative, and complex mathematical reasoning as well as precise and appropriate mathematical language throughout the task. Theoretical knowledge is apparent and applied to concrete situations as the student successfully draws conclusions based on the investigations.

Score Point 3: Proficient

The student demonstrates a broad understanding of solving equations by graphing and elimination. The student uses precise and appropriate language most of the time. Theoretical knowledge is apparent and applied to concrete situations as the student attempts to draw conclusions based on the investigations.

Score Point 2: Apprentice

The student demonstrates an understanding of solving equations by graphing and elimination. The student uses mathematical reasoning most of the time as well as appropriate mathematical language some of the time and attempts to apply theoretical knowledge to the task, but may not be able to apply conclusions based on the investigations.

Score Point 1: Novice

The student demonstrates some understanding of solving equations by graphing and elimination but is unable to complete the task. The student uses little mathematical reasoning or appropriate mathematical language. Theoretical knowledge may appear weak and many responses may be illogical as directions are followed incorrectly.

Answers

Score Point 0: Unsatisfactory

The student fails to make an attempt to complete the task and the responses only restate the problem.

Form B

1.–4. Answers will vary.

Score Point 4: Distinguished

The student demonstrates a comprehensive understanding of formulating problems given specific parameters. The student uses perceptive, creative, and complex mathematical reasoning as well as precise and appropriate mathematical language throughout the task. Theoretical knowledge is apparent and applied to concrete situations as the student successfully draws conclusions based on the investigations.

Score Point 3: Proficient

The student demonstrates a broad understanding of formulating problems given specific parameters. The student uses precise and appropriate language most of the time. Theoretical knowledge is apparent and applied to concrete situations as the student attempts to draw conclusions based on the investigations.

Score Point 2: Apprentice

The student demonstrates an understanding of formulating problems given specific parameters. The student uses mathematical reasoning most of the time as well as appropriate mathematical language some of the time and attempts to apply theoretical knowledge to the task, but may not be able to apply conclusions based on the investigations.

Score Point 1: Novice

The student demonstrates some understanding of formulating problems given specific parameters but is unable to complete the task. The student uses little mathematical reasoning or appropriate mathematical language. Theoretical knowledge may appear weak and many responses may be illogical as directions are followed incorrectly.

Score Point 0: Unsatisfactory

The student fails to make an attempt to complete the task and the responses only restate the problem.

Chapter 8

Quick Warm-Up 8.1

1. 16   2. 24   3. -24 4. -24 5. 24
6. 98

Lesson Quiz 8.1

1. 125   2. 16   3. 49   4. 100   5. 32
6. 36   7. 729   8. 243   9. 12   10. 256
11. 10,000   12. 1   13. 10^8 14. 2^6
15. 5^5 16. 4^8 17. 10^7 18. 7^9 19. $12a^5$
20. $-2b^6$ 21. $18m^4n^2$ 22. $10r^5s$
23. x^5y^7 24. $32a^3b^6c^5$ 25. $6x^2y^2$

Quick Warm-Up 8.2

1. 49   2. $\frac{1}{9}$ 3. 80   4. 64   5. 32

Lesson Quiz 8.2

1. 4096   2. 1,000,000   3. 4096   4. x^9
5. $-y^6$ 6. $-125t^3$ 7. b^6 8. $2s^2$ 9. $25m^4$

Answers

10. $9a^2b^6$ 11. $16r^8s^4$ 12. $-27p^3t^3$
13. $16c^4d^2$ 14. $4m^6n^3$ 15. $-y^4z^6$
16. $-125c^3t^6$ 17. $64e^8f^{10}$ 18. $5g^9h^{15}$
19. a^8 20. a^3b^{10} 21. $c^{11}d^7$ 22. $p^{16}q^8$
23. $5b^8c^8$ 24. $-108r^{13}s^8$ 25. c^4d^6

Quick Warm-Up 8.3

1. -9 2. 6   3. 32   4. 125   5. m^{18}
6. y^{4r} 7. $243a^{10}$ 8. p^4q^6

Lesson Quiz 8.3

1. 5^4; 625   2. 3^1; 3   3. 2^3; 8   4. m^5 5. $9c^3$
6. a^{p-q} 7. $\dfrac{r^3}{s^3}$ 8. $\dfrac{25c^2}{d^4}$ 9. $9g^{10}$ 10. n^3p^3
11. $\dfrac{d^2z^6}{4}$ 12. $\dfrac{j^{2w}k^{5w}}{3^w}$ 13. $5c^5$ 14. $-\dfrac{t^3s^4}{3}$
15. $\dfrac{3uv^7}{4}$ 16. $0.4a^5b^2$ 17. $\dfrac{d^2e^6}{6.25}$ 18. $-b^3d^3$

Quick Warm-Up 8.4

1. 729   2. 128   3. 81   4. 64   5. k^{12}
6. $-8t^3$ 7. a^3b 8. $4v^6$

Lesson Quiz 8.4

1. a   2. c   3. d   4. b   5. $-\dfrac{1}{9}$ 6. $\dfrac{x^2}{y^3}$
7. $\dfrac{b^4}{a^5}$ 8. $\dfrac{1}{c}$ 9. 1   10. $\dfrac{3}{t^3}$ 11. $5m^6$ 12. $\dfrac{r}{2}$
13. $10x^3$ 14. $-\dfrac{1}{a^4}$ 15. $\dfrac{1}{10^7}$ 16. m^7
17. 0   18. 1   19. $\dfrac{1}{37.89}$

Mid-Chapter Assessment—Chapter 8

1. c   2. a   3. d   4. a   5. b   6. 3
7. 100,000   8. $80a^{10}$ 9. $\dfrac{y^5}{5x}$

10. When you raise a power to a power, you multiply the exponents. Multiplication is commutative: $a \cdot b = b \cdot a.$

Quick Warm-Up 8.5

1. 625   2. $\dfrac{1}{100}$, or 0.01   3. 10,000   4. $\dfrac{1}{9}$
5. 320   6. 0.032

Lesson Quiz 8.5

1. The value of the number becomes 100 million times larger.
2. 3×10^6 3. 7.2×10^5 4. 6.34×10^7
5. 8×10^{-3} 6. 7.5×10^{-5}
7. 9.05×10^{-7} 8. 500,000
9. 0.00000007   10. 0.83   11. 60,200,000
12. 29,070,000,000   13. 0.0000015
14. 8×10^8 15. 2.1×10^{12}
16. 4.5×10^{-6} 17. 2×10^3
18. 4.5×10^{-5} 19. 5×10^6

Quick Warm-Up 8.6

1. 1   2. 3   3. 9   4. 27   5. 81
6. 243   7. 1   8. 0.5   9. 0.25
10. 0.125   11. 0.0625   12. 0.03125

Lesson Quiz 8.6

1. $P = A(1 + r)^t$, where P is the amount after t years at a growth rate of r, expressed as a decimal, and A is the original amount.

2.
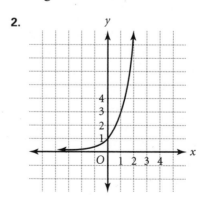

Answers

3.

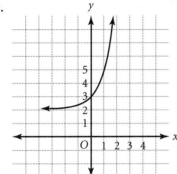

4.

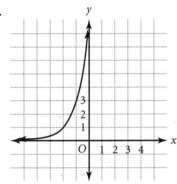

5. $y = 9$ **6.** $y = 1$ **7.** $y = \frac{1}{3}$ **8.** $y = \frac{1}{9}$

9. 58 million

10. 59.4 million

Quick Warm-Up 8.7

1. 64   **2.** 81   **3.** 48   **4.** 252

5. 2916

Lesson Quiz 8.7

1. $P = A(1 + r)^t$ **2.** $P = A(1 - r)^t$

3. \$1050   **4.** \$1102.50   **5.** \$1157.63

6. \$1215.51   **7.** \$1276.28   **8.** $1000(1.05)^x$

9. 10,267,101   **10.** 9,963,961

Chapter Assessment—Form A— Chapter 8

1. d   **2.** a   **3.** b   **4.** d   **5.** c   **6.** b   **7.** a

8. c   **9.** b   **10.** d   **11.** b   **12.** a   **13.** c

14. b   **15.** c

Form B

1. 64   **2.** 1,000,000   **3.** 1.5   **4.** $30p^{11}$

5. $21a^8b^3$ **6.** $-\dfrac{c^3}{2d^5}$ **7.** $75x^2y$ **8.** $\dfrac{6}{r^3}$

9. $-\dfrac{3w^3}{5v^2}$ **10.** $\dfrac{a^2}{2b^5}$ **11.** $-\dfrac{x}{y^4}$

12. Answers may vary. Sample answer: Each number inside the parentheses is raised to the power. Thus, $(5c)^{-3} = (5^{-3})(c^{-3}) = \dfrac{1}{125c^3}$.

13. 308,000   **14.** -0.0002 **15.** 9×10^7

16. 3.05×10^2 **17.** 7×10^{-5}

18. 4.6×10^{-4} **19.** 1.28×10^7

20. 6.09×10^{-6} **21.** 8,000,000   **22.** 720,000

23. 10,500,000   **24.** 0.00003

25. 0.00000000206   **26.** 0.00051

27. 6×10^8 **28.** 3.5×10^{15} **29.** 3.5×10^5

30. 5×10^6 **31.** 2×10^6 **32.** 4×10^3

33.

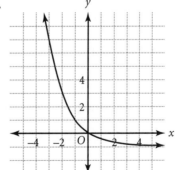

Answers

34.

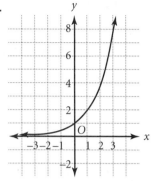

35. $64,405   **36.** $101,705   **37.** $16,831

38. $6100   **39.** 130,000,000

40. 143,000,000

Alternative Assessment—Form A—Chapter 8

1. The Quotient-of-Powers Property, $\frac{x^a}{x^b} = x^{a-b}$, can be used to simplify the quotient. The quotient is $2ab^2$.

2. Multiply the coefficients. Then group powers of like bases. The product is $-5x^3y^2$.

3. Apply the exponent outside the parentheses to each factor of the monomial inside the parentheses; $(-2m^2)^3 = (-2)^3(m^2)^3 = -8m^6$

4. $(a^2b^{-1})^0 = a^0b^0 = 1 \cdot 1 = 1$

5. $-(rt)^2 = -r^2t^2$ and $(-rt)^2 = r^2t^2$; they are opposites.

Score Point 4: Distinguished

The student demonstrates a comprehensive understanding of exponents and powers of monomials. The student uses perceptive, creative, and complex mathematical reasoning as well as precise and appropriate mathematical language throughout the task. Theoretical knowledge is apparent and applied to concrete situations as the student successfully draws conclusions based on the investigations.

Score Point 3: Proficient

The student demonstrates a broad understanding of exponents and powers of monomials. The student uses precise and appropriate language most of the time. Theoretical knowledge is apparent and applied to concrete situations as the student attempts to draw conclusions based on the investigations.

Score Point 2: Apprentice

The student demonstrates an understanding of exponents and powers of monomials. The student uses mathematical reasoning most of the time as well as appropriate mathematical language some of the time and attempts to apply theoretical knowledge to the task, but may not be able to apply conclusions based on the investigations.

Score Point 1: Novice

The student demonstrates some understanding of exponents and powers of monomials but is unable to complete the task. The student uses little mathematical reasoning or appropriate mathematical language. Theoretical knowledge may appear weak and many responses may be illogical as directions are followed incorrectly.

Score Point 0: Unsatisfactory

The student fails to make an attempt to complete the task and the responses only restate the problem.

Form B

1. *A* represents the number of words that Althea memorizes; *r* represents the

Answers

decreasing rate of growth; t represents the time in days; and y represents the number of words that she remembers after t days.

2. $A = 50$, $r = 10\%$, and $t =$ time in days; thus, $y = 50(1 - 0.1)^t$.

3.

d	0	1	2	3	4
y	50	45	40.5	36.5	29.5

d	5	6	7	8	9
y	29.5	26.6	24.9	21.5	19.4

d	10
y	17.4

4. After 6 days, Althea remembers 26.6 words. After 7 days, Althea remembers 24.9 words. Therefore, Althea forgot about 25 words during the first 7 days.

5. The number of words that Althea remembers is decreasing at a rate of 10%.

Score Point 4: Distinguished

The student demonstrates a comprehensive understanding of using exponential functions to solve problems about growth and decay. The student uses perceptive, creative, and complex mathematical reasoning as well as precise and appropriate mathematical language throughout the task. Theoretical knowledge is apparent and applied to concrete situations as the student successfully draws conclusions based on the investigations.

Score Point 3: Proficient

The student demonstrates a broad understanding using exponential functions to solve problems about growth and decay. The student uses precise and appropriate language most of the time. Theoretical knowledge is apparent and applied to concrete situations as the student attempts to draw conclusions based on the investigations.

Score Point 2: Apprentice

The student demonstrates an understanding of using exponential functions to solve problems about growth and decay. The student uses mathematical reasoning most of the time as well as appropriate mathematical language some of the time and attempts to apply theoretical knowledge to the task, but may not be able to apply conclusions based on the investigations.

Score Point 1: Novice

The student demonstrates some understanding of using exponential functions to solve problems about growth and decay but is unable to complete the task. The student uses little mathematical reasoning or appropriate mathematical language. Theoretical knowledge may appear weak and many responses may be illogical as directions are followed incorrectly.

Score Point 0: Unsatisfactory

The student fails to make an attempt to complete the task and the responses only restate the problem.

Chapter 9

Quick Warm-Up 9.1

1. $-b$ 2. $5y$ 3. $3m + 7$ 4. $-2c - 1$

5. $r + 12$ 6. $-4g$ 7. $j - k$

Answers

Lesson Quiz 9.1

1. cubic trinomial

2. $5x^2 + 5x - 8$

3. $3x^3 + 4x^2 - 3x - 5$

4. $-5y^3 + 3y^2 + 4y - 7$

5. $6t^4 + 5t^3 - 5t^2 + 3$

6. $-7a^3 + a^2 + 5a - 4$

7. $-3p^3 - 5p^2 - 2$

8. $r^3 - 4r^2 + 14r - 11$

9. $-13y^3 - 7y^2 + 7y + 5$

10. $7y^2 + 5y + 6$

Quick Warm-Up 9.2

1. 14   2. $-12t$ 3. $3s + 15$ 4. $-4n + 8$

5. $-h + 2$

Lesson Quiz 9.2

1. c

2. $x^2 - 4$

3. $y^2 - 8y + 16$

4. $4m^2 + 12m + 9$

5. $d^2 - 64$

6. $9c^2 - 1$

7. $25r^2 + 40r + 16$

8. $2x - 10$

9. $2r^2 - 2r$

10. $-5g - 10$

11. $-s^2 - 2s$

Quick Warm-Up 9.3

1. $24z^2$

2. $-2w - 10$

3. $q - 9$

4. $2k^2 - 9k + 9$

5. $d^2 - 1$

Lesson Quiz 9.3

1. $(x + a)(x + b) = x(x + b) + a(x + b)$
   $= x^2 + bx + ax + ab$

2. no; $(x + 2)^2$ means $(x + 2)(x + 2)$ which
   is $x^2 + 2x + 4$.

3. $x^2 + 10x + 21$

4. $y^2 + 7y - 60$

5. $r^2 - 13r + 40$

6. $a^2 - 5a - 36$

7. $2b^2 + b - 3$

8. $3t^2 - 11t + 6$

9. $2p^2 + 8p - 10$

10. $x^2 - x + \frac{1}{4}$

11. $2y^2 - \frac{1}{2}y - \frac{1}{4}$

12. $6z^2 + 11z - 10$

13.

14. $y^2 + 3y - 10$

Answers

Quick Warm-Up 9.4

1.

$y = 2x + 1$	
x	y
-3	-5
0	1
1	3
7	15

2.

$y = 7 - x$	
x	y
-6	13
-4	11
2	5
8	-1

Lesson Quiz 9.4

1. 3600 cubic centimeters
2. 1860 square centimeters
3. 7200 cubic centimeters
4. 1530 square centimeters
5. $V = 600x$
6. $1, 0, 1, 4, 9$
7. $-4, -6, -6, -4, 0$
8. $18, 10, 4, 0, -2$
9. $0, 4, 10, 18, 28$
10. $0, -8, -14, -18, -20$

Mid-Chapter Assessment—Chapter 9

1. a   2. c   3. d   4. c   5. a   6. d
7. $7x^4 + 3x^2 - x + 3$
8. $2x^3 + 10x^2 - 5x + 4$
9. $21x^4 - 15x^3 + 6x^2$
10. $12x^2 + 2x - 2$
11. $V = y^3$
12. $0, -4, -6, -6, -4$

Quick Warm-Up 9.5

1. 3   2. 4   3. 1   4. $4h^2 - 20$
5. $2b^3 - 18b^2$

Lesson Quiz 9.5

1. 25
2. r^2t
3. $3xy^2$
4. $a + b$
5. $3(x^2 - 3)$
6. $5x(x + 3)$
7. $y^3(y^2 - 1)$
8. $6t(t^3 + 3t - 4)$
9. $4(7 + 4n^3 - 12n)$
10. $ab(3a^2b - 15a + 20b^2)$
11. $(x + 5)(x - 3)$
12. $(a - b)(x - 2y)$
13. $(2n + 3)(5 + m)$
14. $(3 + x)(x - 1)$
15. $(pq + st)(a - b)^n$
16. $(a - 5)(b + c)$
17. $3(4y^2z - 3)(1 + 2y^3z^2)$

Quick Warm-Up 9.6

1. $4j + 24$
2. $n^2 - 81$
3. $4t^2 + 20t + 25$
4. $3(k^2 + 7)$
5. $y(2y - 15)$
6. $3c^2(2c + 3)$

Lesson Quiz 9.6

1. $x^2 + 4x + 4$
2. $y^2 - 9$

Answers

3. $25a^2 - 40a + 16$

4. $4a^2 - 25b^2$

5. 25 6. $16y$ 7. $12a$ 8. 25

9. $(y - 9)(y + 9)$

10. $(7a - 5)(7a + 5)$

11. $(b + 8)^2$

12. $(m - 3)^2$

13. $4(x - 2)^2$

14. $9(2p + 1)^2$

15. $(c + d)(c - d)(c^2 + d^2)$

16. $x^2(5x - 2)(5x + 2)$

17. $(3a + b)^2$

18. $(y - 2)(y + 2)(y - 3)(y + 3)$

19. $a - 7$

20. $4a - 28$

Quick Warm-Up 9.7

1. $8(3v + 2)$

2. $6t(t - 3)$

3. $2(4a^4 + 8a^2 + 5)$

4. $(n + 12)(n - 12)$

5. $(4z - 5)^2$

Lesson Quiz 9.7

1. 1, 24; 2, 12; 3, 8; 4, 6

2. 1, 10; (2, 5)

3. 1, 12; 2, 6; (3, 4)

4. 1, 18; (2, 9); 3, 6

5. 1, 30; 2, 15; (3, 10); 5, 6

6. $(x + 4)(x - 1)$

7. $2(x + 1)(x + 3)$

8. $(x - 5)(x - 1)$

9. $(x - 4)(x + 3)$

10. $(y - 6)(y + 6)$

11. $5x(2x - 1)$

12. $a^2(3a + 4)(3a - 4)$

13. $12b^2(3 - b^3)$

14. $(d + 9)(d - 2)$

15. $(x - 6)(x + 4)$

Quick Warm-Up 9.8

1. $(u + 2)(u + 4)$

2. $(k + 2)^2$

3. $(b + 3)(b - 4)$

4. $j = -10$

5. $c = 14$

6. $x = -10$

Lesson Quiz 9.8

1. -3 and 1

2. -5 and 5

3. -4 and -1

4. 0 and 6

5. -2 and -3

6. 5

7. $x = -1$ and $x = 2$

8. $x = -2$

9. $x = -4$ and $x = 3$

Answers

10. $x = -5$ and $x = 6$

11. $x = -1$ and $x = 4$

12. $x = -\frac{1}{2}$ and $x = 3$

13. $x = \frac{1}{2}$ and $x = 1$

14. $x = -1$ and $x = 1$

15. $x = -\frac{2}{3}$ and $x = 1$

16. $x = \frac{1}{2}$

Chapter Assessment—Form A—Chapter 9

1. d   2. b   3. a   4. a   5. c   6. d   7. a

8. b   9. a   10. d   11. d   12. c   13. b

14. b   15. a   16. c

Form B

1. cubic binomial

2. There are exactly two factors, 1 and $3x + 4y$.

3. Answers may vary. Sample answer: $x^2 + 6x + 9$

4. Same; the sign of the last term is positive.

5. Negative; the sign of the middle term is negative.

6. $12y^3 - y^2 + 2y - 15$

7. $-w^3 + 15w - 8$

8. $6a^4 - 2a^3 - 2a^2 + 7a + 7$

9. $3x^3 - 12x^2 + 5x + 12$

10. $-b^3 - 9b^2 + 3b - 13$

11. $5d^4 + 3d^3 - 2d^2 - 5d - 2$

12. $-4x^3 + x + 11$

13. $12y^3 - 12y^2 - 1$

14. $3x^2 - 5x - 12$

15. $15w^3 - 3w$

16. $16a^2 - 4b^4$

17. $x^2 + 6x + 9$

18. $4x^2 - 12x + 9$

19. $d^2 + 2d - 15$

20. $6t^2 - 7t - 24$

21. $(3r - 2)(3r + 2)$

22. $(z + 9)(z - 2)$

23. $(m - 5)^2$

24. $7a(a + 5)(a - 1)$

25. $(y - 6)(y + 3)$

26. $(b + 5)(b - 1)$

27. $x = -4$ and $x = 1$

28. $x = -2$ and $x = 2$

29. $x = 3$ and $x = 5$

30. $x = 1$

31. $10x + 20$

32. $6x^2 + 23x + 21$

33. $2x^2 + 23x + 21$

34. $2p - 3$

35. $8p - 12$

Alternative Assessment—Form A—Chapter 9

1. Find the difference by adding the opposite of $3x^2 + 1$ to $4x^2 + 2x - 1$ and combine like terms. The difference is $x^2 + 2x - 2$.

Answers

2.

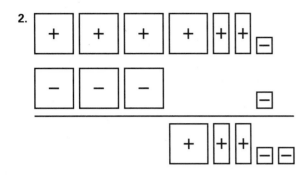

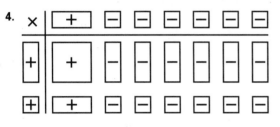

3. Sample answer: Use the Distributive Property. The product is $x^2 - 5x - 6$.

4.

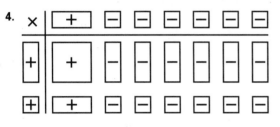

5. $2x^2(x + 2)(x - 2)$

Score Point 4: Distinguished

The student demonstrates a comprehensive understanding of operations with polynomials. The student uses perceptive, creative, and complex mathematical reasoning as well as precise and appropriate mathematical language throughout the task. Theoretical knowledge is apparent and applied to concrete situations as the student successfully draws conclusions based on the investigations.

Score Point 3: Proficient

The student demonstrates a broad understanding of operations with polynomials. The student uses precise and appropriate language most of the time. Theoretical knowledge is apparent and applied to concrete situations as the student attempts to draw conclusions based on the investigations.

Score Point 2: Apprentice

The student demonstrates an understanding of operations with polynomials. The student uses mathematical reasoning most of the time as well as appropriate mathematical language some of the time and attempts to apply theoretical knowledge to the task, but may not be able to apply conclusions based on the investigations.

Score Point 1: Novice

The student demonstrates some understanding of operations with polynomials but is unable to complete the task. The student uses little mathematical reasoning or appropriate mathematical language. Theoretical knowledge may appear weak and many responses may be illogical as directions are followed incorrectly.

Score Point 0: Unsatisfactory

The student fails to make an attempt to complete the task and the responses only restate the problem.

Form B

1. Answers may vary. Sample answer: Try to find a factor that you can multiply by $3x - 1$ in order to get the product $3x^2 - 4x + 1$; Since $(3x - 1)(x - 1) = 3x^2 - 4x + 1$, the second factor is $x - 1$.

2. $2x(x^2 - 2x + 4)$.

3. prime

4. $y(10x + y)$

5. The polynomial is prime if it has exactly two factors, 1 and itself.

6. $(5n - 1)(n + 1)$

Answers

7. $25(2x + 1)(2x - 1)$

8. $(3p + 1)^2$

9. Multiply the factors.

Score Point 4: Distinguished

The student demonstrates a comprehensive understanding of factoring polynomials. The student uses perceptive, creative, and complex mathematical reasoning as well as precise and appropriate mathematical language throughout the task. Theoretical knowledge is apparent and applied to concrete situations as the student successfully draws conclusions based on the investigations.

Score Point 3: Proficient

The student demonstrates a broad understanding of factoring polynomials. The student uses precise and appropriate language most of the time. Theoretical knowledge is apparent and applied to concrete situations as the student attempts to draw conclusions based on the investigations.

Score Point 2: Apprentice

The student demonstrates an understanding of factoring polynomials. The student uses mathematical reasoning most of the time as well as appropriate mathematical language some of the time and attempts to apply theoretical knowledge to the task, but may not be able to apply conclusions based on the investigations.

Score Point 1: Novice

The student demonstrates some understanding of factoring polynomials but is unable to complete the task. The student

uses little mathematical reasoning or appropriate mathematical language. Theoretical knowledge may appear weak and many responses may be illogical as directions are followed incorrectly.

Score Point 0: Unsatisfactory

The student fails to make an attempt to complete the task and the responses only restate the problem.

Chapter 10

Quick Warm-Up 10.1

1. $x = -6$ or $x = 2$ 2. $x = -5$ or $x = 6$

3. $x = 4$ 4. $x = -1$ or $x = 1$

5. $x = 0$ or $x = 1$

Lesson Quiz 10.1

1. The parent function moves 4 units to the right and 1 unit up.

2. The parent function is stretched by a factor of $\frac{1}{2}$ and shifted 3 units down.

3. The parent function is stretched by a factor of 3 and shifted 2 units to the left.

4. The parent function is reflected through the x-axis and shifted 1 unit to the right and 2 units up.

5. vertex: $(-3, -1)$; axis: $x = -3$

6. vertex: $(5, 0)$; axis: $x = 5$

Answers

7.

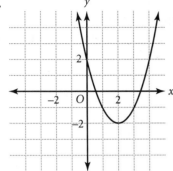

8.

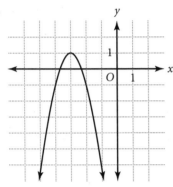

Quick Warm-Up 10.2

1. 25   2. 121   3. $x = 8$ 4. $y = -4$

5. $t = 20$

Lesson Quiz 10.2

1. 68, 28, −28

2. about $(3.5, 0)$

3. 13

4. 9

5. 15

6. 4.12

7. $x = -7$ and $x = 7$

8. $x = -0.67$ and $x = 0.67$

9. $x = 2$ and $x = -6$

10. $x = -1$ and $x = 11$

11. $x = 5.65$ and $x = 0.35$

12. $x = -9$ and $x = 1$

13. $(-5, -4)$

14. $x = -5$

15. -7 and -3

16.

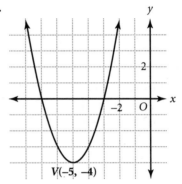

$V(-5, -4)$

Quick Warm-Up 10.3

1. 9   2. 25   3. $\dfrac{49}{4}$ 4. $\dfrac{1}{4}$ 5. $(x + 1)^2$

6. $(x + 4)^2$ 7. $(x - 5)^2$

Lesson Quiz 10.3

1. (h, k)

2. 16; $(x + 4)^2$

3. 4; $(x - 2)^2$

4. 49; $(x - 7)^2$

5. $y = (x + 6)^2 - 36$

6. $y = (x - 9)^2 - 81$

7. $y = (x + 8)^2 - 64$

8. $(-8, -64)$

9. The minimum value is −64.

10. $y = (x - 2)^2 - 1$; vertex: $(2, -1)$

11. $y = \left(x + \dfrac{3}{2}\right)^2 + \dfrac{11}{4}$; vertex: $\left(-\dfrac{3}{2}, \dfrac{11}{4}\right)$

Answers

Mid-Chapter Assessment—Chapter 10

1. d   2. b   3. c   4. a   5. a

6. $x = -15$ and $x = 15$

7. $x = 7$ and $x = -15$

8. $x = -4.24$ and $x = 0.24$

9. $x = 4$ and $x = 6$

10. $y = (x - 3)^2 - 8$

11. $(3, -8)$

12. $x = 3$

13. 5.83 and 0.17

14.

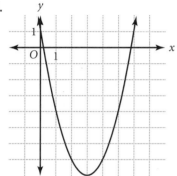

Quick Warm-Up 10.4

1. $x = -1$ or $x = 7$ 2. $x = 1$ or $x = -9$

3. $(x + 3)(x - 2)$ 4. $(x - 2)(x - 6)$

Lesson Quiz 10.4

1. $x = 4$ and $x = 6$

2. $x = -5$ and $x = -6$

3. $x = -2$ and $x = 9$

4. $x = -7$ and $x = 3$

5. $x = -6$ and $x = 0$

6. $x = -0.24$ and $x = 4.24$

7. $x = 0.53$ and $x = 9.47$

8. $x = -3.83$ and $x = 1.83$

9.

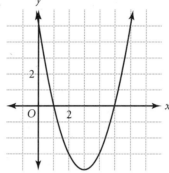

10. $x = 7$ and $x = -1$

11. $x = 3$

12. $x = 0$ and $x = 6$

13. $x = 5$ and $x = 1$

Quick Warm-Up 10.5

1. 8   2. ± 12 3. 24   4. 16   5. -1

Lesson Quiz 10.5

1. $a = 1, b = -7, c = 11$

2. $a = 3, b = 2, c = -8$

3. -23; no real solutions

4. 196; two real solutions

5. 25; two real solutions

6. $x = -6$ and $x = -1$

7. $y = -\frac{1}{2}$ and $y = -3$

8. $x = 6$ and $x = -1$

9. $r = -6.46$ and $r = 0.46$

10. $y = -3$ and $y = -5$

11. $y = -5$ and $y = 5$

12. $x = 5.83$ and $x = 0.17$

13. $t = 0$ and $t = 6$

Answers

Quick Warm-Up 10.6

1. $x = -3$ or $x = 5$ 2. $x = -7$ or $x = 4$

3. yes   4. yes   5. no

Lesson Quiz 10.6

1. $x > -2$ or $x < -5$

2. $-5 < x < 6$

3. $x \geq 2$ or $x \leq 6$

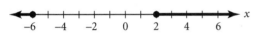

4. $-5 \leq x \leq 4$

5.

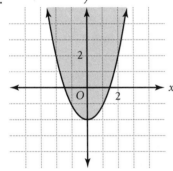

6.

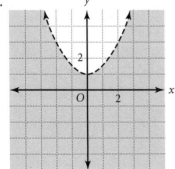

Chapter Assessment—Form A—Chapter 10

1. d   2. a   3. c   4. c   5. b   6. a   7. d

8. c   9. c   10. c   11. d   12. c   13. a

14. a   15. c   16. c

Form B

1. vertex: $(-1, -3)$; axis of symmetry: $x = -1$

2. vertex: $(4, -5)$; axis of symmetry: $x = 4$

3. $(x - 5)(x - 3)$

4. $(x + 1)(x + 3)$

5. $(x + 8)(x - 3)$

6. $3(x - 1)(x - 2)$

7. 500 feet   8. 5 seconds   9. 10 seconds

10. $x = 5$ 11. 15   12. 11   13. 4.80

14. 9.75

15. $x = -14$ and $x = 14$

16. $x = -2$ and $x = 8$

17. $x = -6$ and $x = -4$

18. $x = -2.12$ and $x = 6.12$

19. $y = (x + 3)^2 - 18$; vertex: $(-3, -18)$

20. $y = \left(x - \frac{5}{2}\right)^2 - \frac{11}{2}$; vertex: $\left(\frac{5}{2}, -\frac{11}{2}\right)$

21. $x = -2$ and $x = 10$

22. $t = -6$ and $t = 0$

23. $w = -5$ and $w = 5$

24. $y = 3$ and $y = \frac{5}{2}$

25. $x = -0.57$ and $x = 2.91$

26. $r = -2.81$ and $r = 0.21$

27. $(1, 0)$ and $(3, 4)$

Answers

28. $(2, 7)$ and $(-7, -2)$

29. $x \geq 2$ or $x \leq -5$

30. $1 < x < 4$

31.

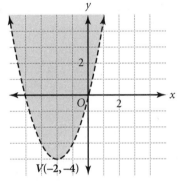

32.

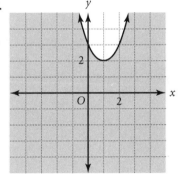

Alternative Assessment—Form A— Chapter 10

1. 10; Since the second differences are constant, the data represents a quadratic polynomial.

2. The x-values are -4 and 10. The negative value means that Carmen would be paying customers to buy a teddy bear. The positive value means that if Carmen decreases the price by $10, her total income would be $0.

3. To determine the axis of symmetry, find the average of the x-values found in

Exercise 2. The axis of symmetry is

$$x = \frac{(-4 + 10)}{2} = 3.$$

4. Use the table of values to find the corresponding y-value for $x = 3$. The vertex is $(3, 245)$; the maximum value is 245.

Score Point 4: Distinguished

The student demonstrates a comprehensive understanding of quadratic functions. The student uses perceptive, creative, and complex mathematical reasoning as well as precise and appropriate mathematical language throughout the task. Theoretical knowledge is apparent and applied to concrete situations as the student successfully draws conclusions based on the investigations.

Score Point 3: Proficient

The student demonstrates a broad understanding of quadratic functions. The student uses precise and appropriate language most of the time. Theoretical knowledge is apparent and applied to concrete situations as the student attempts to draw conclusions based on the investigations.

Score Point 2: Apprentice

The student demonstrates an understanding of quadratic functions. The student uses mathematical reasoning most of the time as well as appropriate mathematical language some of the time and attempts to apply theoretical knowledge to the task, but may not be able to apply conclusions based on the investigations.

Score Point 1: Novice

The student demonstrates some understanding of quadratic functions but is

Answers

unable to complete the task. The student uses little mathematical reasoning or appropriate mathematical language. Theoretical knowledge may appear weak and many responses may be illogical as directions are followed incorrectly.

Score Point 0: Unsatisfactory

The student fails to make an attempt to complete the task and the responses only restate the problem.

Form B

1. Answers may vary. Sample answer: Find the number that must be added in order to complete the square. Add and subtract this number and group like terms. Write in the form
$y = (x - h)^2 + k$. $(n + 1)^2 - 25 = 0$; $n = 4$ and $n = -6$

2. $x = -\frac{5}{2}$ and $x = 3$; to check, substitute each solution into the original equation.

3. Answers may vary. Sample answer: Write the equation in the form $ax^2 + bx + c = 0$. Then substitute the values for a, b, and c into the quadratic formula. The solutions are 0.5 and -3.5.

4. Answers may vary. Sample answer: The quadratic formula is the most direct method because the leading coefficient is 4. The solutions are $\frac{1}{2}$ and $-\frac{7}{2}$.

Score Point 4: Distinguished

The student demonstrates a comprehensive understanding of quadratic equations. The student uses perceptive, creative, and complex mathematical reasoning as well as precise and appropriate mathematical language

throughout the task. Theoretical knowledge is apparent and applied to concrete situations as the student successfully draws conclusions based on the investigations.

Score Point 3: Proficient

The student demonstrates a broad understanding of quadratic equations. The student uses precise and appropriate language most of the time. Theoretical knowledge is apparent and applied to concrete situations as the student attempts to draw conclusions based on the investigations.

Score Point 2: Apprentice

The student demonstrates an understanding of quadratic equations. The student uses mathematical reasoning most of the time as well as appropriate mathematical language some of the time and attempts to apply theoretical knowledge to the task, but may not be able to apply conclusions based on the investigations.

Score Point 1: Novice

The student demonstrates some understanding of quadratic equations but is unable to complete the task. The student uses little mathematical reasoning or appropriate mathematical language. Theoretical knowledge may appear weak and many responses may be illogical as directions are followed incorrectly.

Score Point 0: Unsatisfactory

The student fails to make an attempt to complete the task and the responses only restate the problem.

Answers

Chapter 11

Quick Warm-Up 11.1

1. 216   2. 28.8   3. 18   4. 50.4   5. 14

Lesson Quiz 11.1

1. Not an inverse   2. Inverse   3. Inverse

4. Not an inverse   5. $x = 2$ 6. $x = -14$

7. $y = \frac{1}{4}$ 8. $y = 3$ 9. $x = 2$ 10. $y = 1$

11. 500 revolutions per minute

12. $6\frac{1}{4}$ hours

Quick Warm-Up 11.2

1. $y = 3.75$

2. $c = 0$ or $c = 4$

3. $m = 1$

4. $a = -2$ or $a = 9$

Lesson Quiz 11.2

1. $f(x) = \frac{5}{x} + 3$

2. $x = 0$ 3. $x = -1$

4. $x = 0$ and $x = 2$ 5. $x = -2$ and $x = 3$

6. 3; 0   7. 2; undefined

8. $-1; \frac{1}{5}$ 9. $2; -\frac{2}{5}$

10.

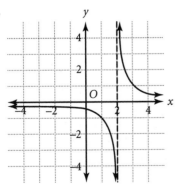

undefined at $x = 2$

11.

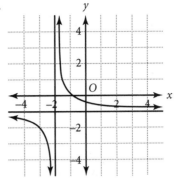

undefined at $x = -2$

Quick Warm-Up 11.3

1. w^3 2. $\frac{xy^3}{5}$ 3. $\frac{1}{a^6}$ 4. $\frac{3m^4}{n^3}$ 5. $5(k - 7)$

6. $(r + 6)(r - 4)$ 7. $(x + 1)(x - 1)$

8. $3(b - 2)(b - 12)$

Lesson Quiz 11.3

1. $2 \cdot 2 \cdot 3 \cdot 5$

2. $(x - 5)(x + 2)$

3. Multiply the expression by -1; $-1(x - 5)$.

4. 12   5. $3a^2b$ 6. d 7. $m(m - 2)$

8. $x + 3$ 9. $y - 2$

10. $\frac{1}{6z^3}, z \neq 0$

11. $\frac{3p^2}{q}, p \neq 0, q \neq 0$

12. $\frac{2x - 3}{2x}, x \neq 0, x \neq -\frac{3}{2}$

13. $\frac{2x}{3}, x \neq 4$

14. $-\frac{3}{a + 4}, a \neq \pm 4$

15. $\frac{m + 1}{m - 1}, m \neq \pm 1$

16. $\frac{y - 3}{2y}, y \neq 0, y \neq -4$

17. $-\frac{5b}{b + 5}, b \neq -5, b \neq 2$

18. $x - 2$

Answers

Quick Warm-Up 11.4

1. $\frac{2}{3}$ 2. $\frac{7}{6}$, or $1\frac{1}{6}$ 3. $\frac{3}{20}$ 4. $\frac{4}{15}$ 5. $\frac{9}{16}$

6. $\frac{1}{10}$

Lesson Quiz 11.4

1. $6xy$ 2. Multiply by $\frac{3x}{3x}$.

3. Divide numerator and denominator by $2xy$.

4. $\frac{31}{10a}$, $a \neq 0$

5. $\frac{4 - 15b}{6b^2}$, $b \neq 0$

6. $\frac{4}{5x^2y}$, $x \neq 0$, $y \neq 0$

7. $\frac{1}{5m}$, $m \neq 0$

8. $\frac{4r - 6}{r - 2}$, $r \neq 2$

9. $\frac{16}{t - 3}$, $t \neq 3$

10. 9, $n \neq 2$

11. $\frac{p}{p - 5}$, $p \neq 0$, $p \neq -5$, $p \neq 5$

12. $\frac{x}{x + 1}$, $x \neq -1$, $x \neq 3$

13. $\frac{w^2 - 2w + 2}{4w^2(w - 1)}$, $w \neq 0$, $w \neq 1$

14. $\frac{z + 3}{4}$, $z \neq 3$, $z \neq -3$

15. -3, $d \neq 1$, $d \neq -1$

Mid-Chapter Assessment—Chapter 11

1. c   2. d   3. b   4. a

5. $\frac{a}{4b^2}$, $a \neq 0$, $b \neq 0$

6. $\frac{3(x + y)}{x - y}$, $x \neq y$

7. $\frac{2m}{m + 3}$, $m \neq 3$, $m \neq -3$

8. $\frac{p + 4}{p + 3}$, $p \neq 2$, $p \neq -3$

9. $\frac{15a - 2b}{3a^2b^2}$, $a \neq 0$, $b \neq 0$

10. $\frac{9}{x - 2}$, $x \neq 2$

11. $\frac{r - 1}{2}$, $r \neq 0$, $r \neq -3$, $r \neq -1$

12. $\frac{w - 3}{16w}$, $w \neq 0$, $w \neq 3$

Quick Warm-Up 11.5

1. $g = 6$ 2. $k = -24$ 3. $n = 3$

4. $t = 2$ or $t = -5$

5. $\frac{x - 2}{6}$, where $x \neq 0$ and $x \neq -2$

Lesson Quiz 11.5

1. $y_1 = \frac{2}{3}x + \frac{1}{x}$ and $y_2 = 5$

2. $\frac{81}{20}$ or $4\frac{1}{20}$

3. 7

4. -12

5. -2 and 12

6.
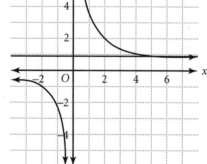

6

Answers

7.

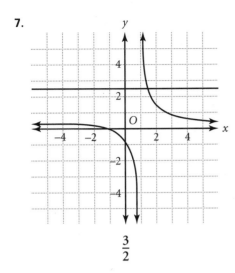

$$\frac{3}{2}$$

Quick Warm-Up 11.6

1. Distributive Property

2. Mult. Property of Zero

3. Addition Prop. of Equality

4. Transitive Prop. of Equality

Lesson Quiz 11.6

1. $\frac{a}{b} = \frac{c}{d}$

2. If $a < 8$, then $a - 5 < 3$.

3. $5m + 3m = (5 + 3)m$ Distributive Prop.
   $= 8m$ Definition of
   Addition

4. $4\left(x - \frac{1}{2}\right) = 22$ Given

   $4x - 2 = 22$ Distributive Prop
   $4x - 2 + 2 = 22 + 2$ Addition Prop.
                          of Equality
   $4x = 24$ Simplify.
   $\frac{4x}{4} = \frac{24}{4}$ Division Prop.
                         of Equality
   $x = 6$ Simplify.

5. Given

6. Subtraction Property of Equality

7. $\frac{b}{b} = 1$ and $\frac{d}{d} = 1$; subtraction

8. Definition of subtraction for rational numbers

Chapter Assessment—Form A—Chapter 11

1. c   2. b   3. d   4. a   5. c   6. a   7. c

8. c   9. b   10. b   11. a   12. d   13. c

14. b   15. b   16. c   17. d   18. a   19. d

Form B

1. A rational expression is undefined when the denominator is equal to 0.

2. $3 - x$ is the opposite of $x - 3$.

3. $a + 3 > 8$

4.

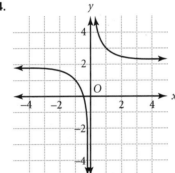

5.

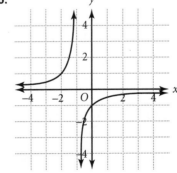

6. $x = 6$ 7. $y = 4$ 8. $x = 8$

9. 640 cycles per second

Answers

10. $\dfrac{2x}{3y}$ 11. $\dfrac{a^2}{a+3}$ 12. 1   13. $\dfrac{3}{4-k}$

14. $\dfrac{5}{n+5}$ 15. $\dfrac{b+3}{b-1}$ 16. $\dfrac{15x+7}{6x^2}$

17. $\dfrac{7a-20}{35a^2}$ 18. $\dfrac{3m^3}{5n}$ 19. $\dfrac{2p^2q}{5}$

20. $\dfrac{(y+3)^2}{2y+3}$ 21. $b-3$ 22. $\dfrac{k^2-2k-3}{6k}$

23. $\dfrac{3z(z-3)}{z+3}$ 24. $x=-4$ 25. $x=\dfrac{2}{3}$

26. $x=1$ 27. $x=-3$ and $x=2$

28. $x=-\dfrac{4}{5}$

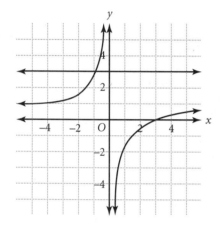

29. 5.8   30. 3   31. 30 minutes

32. The perimeter is $4x+12$.

Alternative Assessment—Form A—Chapter 11

1. Each fraction represents two polynomials, P and Q, in the form $\dfrac{P}{Q}$.

2. The fractions are undefined for $b=0$. Division by 0 is not possible.

3. To find the common denominator, factor the denominators. Then find the common denominator; the sum is $\dfrac{10b+1}{6}$. The difference is $\dfrac{-(2b+1)}{2}$.

4. To multiply rational expressions, multiply the numerators and multiply the denominators; the product $\dfrac{8b^2-2b-1}{18}$ is in simplest form because the only common factor of the numerator and denominator is 1.

5. To divide by a rational expression, multiply by its reciprocal; the quotient is $\dfrac{2b-1}{8b+2}$.

Score Point 4: Distinguished

The student demonstrates a comprehensive understanding of operations with rational expressions. The student uses perceptive, creative, and complex mathematical reasoning as well as precise and appropriate mathematical language throughout the task. Theoretical knowledge is apparent and applied to concrete situations as the student successfully draws conclusions based on the investigations.

Score Point 3: Proficient

The student demonstrates a broad understanding of operations with rational expressions. The student uses precise and appropriate language most of the time. Theoretical knowledge is apparent and applied to concrete situations as the student attempts to draw conclusions based on the investigations.

Score Point 2: Apprentice

The student demonstrates an understanding of operations with rational expressions. The student uses mathematical reasoning most of the time as well as appropriate mathematical language some of the time and attempts to apply theoretical knowledge to the task, but may not be able to apply conclusions based on the investigations.

Answers

Score Point 1: Novice

The student demonstrates some understanding of operations with rational expressions but is unable to complete the task. The student uses little mathematical reasoning or appropriate mathematical language. Theoretical knowledge may appear weak and many responses may be illogical as directions are followed incorrectly.

Score Point 0: Unsatisfactory

The student fails to make an attempt to complete the task and the responses only restate the problem.

Form B

1. Possible answer: Amy and Jan biked in a race to raise money for a class trip. Amy biked 14 miles. Jan biked 9 miles. Jan biked at a rate that was 5 miles per hour slower than Amy. If they both biked the same amount of time, find the rate at which Amy biked.

2. Possible answer: Darius repairs broken in-line skates. He spent $60 to buy the equipment needed to fix the skates. If he now charges $4 per pair, how many pairs would he have to fix before he has an average cost per pair of $9?

3. Possible answer: How many years will it take to double your money at 12%?

Score Point 4: Distinguished

The student demonstrates a comprehensive understanding of problem formation and rational equations. The student uses perceptive, creative, and complex mathematical reasoning as well as precise and appropriate mathematical language throughout the task. Theoretical knowledge is apparent and applied to concrete situations as the student successfully draws conclusions based on the investigations.

Score Point 3: Proficient

The student demonstrates a broad understanding of problem formation and rational equations. The student uses precise and appropriate language most of the time. Theoretical knowledge is apparent and applied to concrete situations as the student attempts to draw conclusions based on the investigations.

Score Point 2: Apprentice

The student demonstrates an understanding of problem formation and rational equations. The student uses mathematical reasoning most of the time as well as appropriate mathematical language some of the time and attempts to apply theoretical knowledge to the task, but may not be able to apply conclusions based on the investigations.

Score Point 1: Novice

The student demonstrates some understanding of problem formation and rational equations but is unable to complete the task. The student uses little mathematical reasoning or appropriate mathematical language. Theoretical knowledge may appear weak and many responses may be illogical as directions are followed incorrectly.

Score Point 0: Unsatisfactory

The student fails to make an attempt to complete the task and the responses only restate the problem.

Answers

Chapter 12

Quick Warm-Up 12.1

1. 196   2. 25   3. 9   4. -1 5. y^2

6. $p^2 - 2p$ 7. $k^2 - 9$ 8. $n^2 - 5n + 4$

Lesson Quiz 12.1

1. 13   2. 7.07   3. 50   4. 6.93   5. 8.49

6. 6.71   7. 7   8. 7.35   9. 8.66   10. 10

11. $6\sqrt{2}$ 12. $4\sqrt{6}$ 13. $\frac{3}{5}$ 14. 4   15. 3

16. ab^2 17. $\frac{m^2\sqrt{m}}{n^2}$ 18. $x^2y^3\sqrt{xy}$

19. $st^3\sqrt{s}$ 20. $c^4\sqrt{b}$ 21. $\frac{g^4\sqrt{gh}}{h^3}$

22. $7\sqrt{3}$ 23. $5\sqrt{2}$ 24. $9\sqrt{3}$ 25. $\sqrt{5}$

26. $2 + \sqrt{5}$ 27. 6

Quick Warm-Up 12.2

1. $m = -1.5$ 2. $s = -6$ or $s = 6$

3. $a = -5$ or $a = 1$ 4. $g = 0$ or $g = -4$

5. $u = -5$ or $u = 1$

Lesson Quiz 12.2

1. $x = 22$ 2. $x = -59$ 3. no solution

4. $x = 22$ 5. 1 and 2   6. $x = 3$

7. $x = 2$ 8. $x = 5$

9. $x = 10$

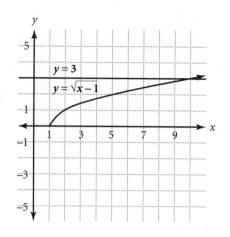

10. $x = 4$

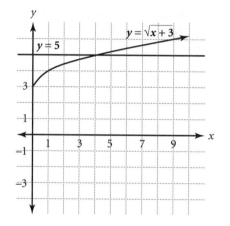

Quick Warm-Up 12.3

1. $x = \pm 7$ 2. $x = \pm 4\sqrt{3}$

3. $x = \pm 5$ 4. $x = \pm 4$

5. $x = \pm\sqrt{y^2 + z^2}$ 6. $x = \pm\sqrt{z^2 - y^2}$

Lesson Quiz 12.3

1. 20   2. 8   3. 5   4. 1   5. 15.6   6. 4.6

7. 11.4   8. 60   9. 16   10. 12   11. 39

12. 15.9   13. 14.4 miles

Mid-Chapter Assessment—Chapter 12

1. d   2. c   3. b   4. a   5. 15   6. $7\sqrt{2}$

7. 300   8. $4\sqrt{3}$ 9. 5   10. $3\sqrt{2}$

Answers

11. $8\sqrt{3}$ 12. $5\sqrt{6}$ 13. $3 + \sqrt{3}$

14. $2 + \sqrt{5}$ 15. ≈ 14.9

16. 4

Quick Warm-Up 12.4

1. 10   2. 3   3. $5\sqrt{2}$ 4. 8   5. 3   6. $2\sqrt{6}$

Lesson Quiz 12.4

1. 5   2. $\sqrt{74}$ 3. $\sqrt{130}$ 4. 10   5. 2

6. $\sqrt{34}$ 7. $(5, 3)$ 8. $(-1, 6)$ 9. $(2, -2)$

10. $(4, -2)$ 11. $B(5, 5)$ 12. $A(-1, 3)$

13. $A(-2, 3), B(2, -5)$ 14. $A(8, -9), B(0, 1)$

15. isosceles; $AB = 5$ and $AC = 5$. Thus, two sides of the triangle are congruent.

Quick Warm-Up 12.5

1. 11   2. 2   3. 5   4. ≈ 9.49 5. $(2, 9)$

6. $(-3, 3)$

Lesson Quiz 12.5

1. $x^2 + y^2 = 16$

2. $x^2 + y^2 = 81$

3. $x^2 + y^2 = 25$

4. $x^2 + y^2 = r^2$

5. center at $(3, 1)$; radius of 5

6. center at $(-5, 0)$; radius of 6

7. center at $(-1, -3)$; radius of 4

8. center at $(0, 2)$; radius of $\sqrt{5}$

Quick Warm-Up 12.6

1. 0.625   2. 0.4   3. $w = 80$ 4. $a = 0.2$

5. 6   6. 4.5

Lesson Quiz 12.6

1. 3.73   2. 0.12   3. 0.58   4. 0.25   5. 0.75

6. 57.29   7. $45°$ 8. $60°$ 9. $12°$ 10. $44°$

11. $56°$ 12. $77°$

13. $\tan A = \frac{12}{5}$; $\tan B = \frac{5}{12}$

14. $\tan X = \frac{8}{15}$; $\tan Y = \frac{15}{8}$

Quick Warm-Up 12.7

1. ≈ 1.5399 2. $\approx 39°$ 3. $\frac{5}{12}$

4. $\frac{12}{5}$ 5. $\approx 23°$ 6. $\approx 67°$

Lesson Quiz 12.7

1. $\frac{8}{17}$ 2. $\frac{8}{17}$ 3. $\frac{8}{15}$ 4. $\frac{15}{17}$

5. $48.2°$ 6. $46.7°$ 7. 12.5   8. 3.1

9. $75°$ 10. $90°$ 11. 0.996   12. 0.574

Quick Warm-Up 12.8

1. -9 2. 1.3   3. -2 4. -0.9

5. -48 6. 63   7. -10 8. 6

Lesson Quiz 12.8

1. 5   2. 6   3. 9   4. -8

5. $\begin{bmatrix} -3.2 & 14 \\ 2 & -4.8 \end{bmatrix}$ 6. $\begin{bmatrix} -6.8 & 6 \\ 2 & -11.2 \end{bmatrix}$

7. $\begin{bmatrix} -4\frac{1}{3} & 1 \\ 8 & -7\frac{1}{2} \end{bmatrix}$ 8. $\begin{bmatrix} 5\frac{2}{3} & -19 \\ 4 & 8\frac{1}{2} \end{bmatrix}$

9. $\begin{bmatrix} 13 & 26 \\ 10 & 113 \\ 18 & 92 \end{bmatrix}$

Answers

Chapter Assessment—Form A—
Chapter 12

1. c   2. b   3. b   4. a   5. c   6. d   7. b

8. d   9. c   10. a   11. b   12. a   13. b

14. d   15. c   16. d   17. d

Form B

1. $x \leq 5$ 2. 9   3. -15 4. 0.5   5. 5.66

6. 7 centimeters

7.
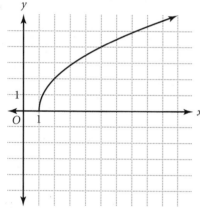

domain: $x \geq 1$; range: $y \geq 0$

8.
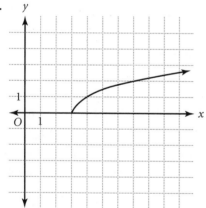

domain: $x \geq 3$; range: $y \geq 0$

9. $6\sqrt{2}$ 10. 0.3   11. $9\sqrt{3}$ 12. $\sqrt{6}$

13. 63   14. $5\sqrt{3}$ 15. $\frac{2\sqrt{21}}{3}$ 16. $-2\sqrt{3}$

17. $3\sqrt{5} - \sqrt{3}$ 18. $x = 52$ 19. $x = 27$

20. $x = 2$ 21. 9   22. 2×4 23. $x = 20$

24. $x = 15$ 25. 10   26. $3\sqrt{5}$

27. right triangle   28. $M(5, 4)$ 29. $M(1, -2)$

30. $D(2, 3)$ 31. $C(8, -2)$

32. center: $(0, 0)$; radius: $\sqrt{5}$

33. center: $(3, -1)$; radius: 7

34.
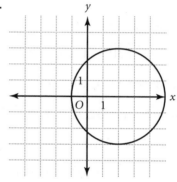

35. 6.67 inches   36. $48°$

Alternative Assessment—Form A—
Chapter 12

1. The statements are false. Sample Answer:
$\sqrt{25} + \sqrt{9} \neq \sqrt{25 + 9}$ and
$\sqrt{25} - \sqrt{9} \neq \sqrt{25 - 9}$

2. The statements are true. Sample Answer:
$\sqrt{25} \cdot \sqrt{9} = \sqrt{25 \cdot 9}$ and $\frac{\sqrt{25}}{\sqrt{9}} = \sqrt{\frac{25}{9}}$

3. $6\sqrt{3} - 2\sqrt{2}$ 4. $2 - \sqrt{2}$

5. The expression is simplified.

6. To add expressions that contain radicals, the radicands must be the same.

7. To simplify radicals, the radicand must be factored so that it contains no perfect squares greater than 1. $\sqrt{98} = \sqrt{49 \cdot 2} = 7\sqrt{2}$.

Answers

8. To multiply differences and sums, use the Distributive Property or the FOIL method. $2\sqrt{6} - 4\sqrt{2} + 3\sqrt{3} - 6$

Score Point 4: Distinguished

The student demonstrates a comprehensive understanding of the properties of radicals. The student uses perceptive, creative, and complex mathematical reasoning as well as precise and appropriate mathematical language throughout the task. Theoretical knowledge is apparent and applied to concrete situations as the student successfully draws conclusions based on the investigations.

Score Point 3: Proficient

The student demonstrates a broad understanding of the properties of radicals. The student uses precise and appropriate language most of the time. Theoretical knowledge is apparent and applied to concrete situations as the student attempts to draw conclusions based on the investigations.

Score Point 2: Apprentice

The student demonstrates an understanding of the properties of radicals. The student uses mathematical reasoning most of the time as well as appropriate mathematical language some of the time and attempts to apply theoretical knowledge to the task, but may not be able to apply conclusions based on the investigations.

Score Point 1: Novice

The student demonstrates some understanding of the properties of radicals but is unable to complete the task. The student uses little mathematical reasoning or appropriate mathematical language. Theoretical knowledge may appear weak and many responses may be illogical as directions are followed incorrectly.

Score Point 0: Unsatisfactory

The student fails to make an attempt to complete the task and the responses only restate the problem.

Form B

1. The diagram is a right triangle. The legs of the triangle are 33 miles and 56 miles.

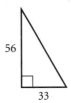

2. The missing length is the hypotenuse of the right triangle.

3. The legs of the right triangle are 33 miles and 56 miles. The broadcasting range is the length of the hypotenuse. Use the Pythagorean Theorem to find the length of the hypotenuse. Use the definition of a square root to solve.

4. The broadcasting range is 65 miles.

5. Check by substitution.

6. The broadcasting range can be modeled by a circle with its center at $(0, 0)$ and a radius of 65 miles. The equation of the circle is $x^2 + y^2 = 4225$.

Score Point 4: Distinguished

The student demonstrates a comprehensive understanding of coordinate geometry. The student uses perceptive, creative, and complex mathematical reasoning as well as precise and appropriate mathematical

Answers

language throughout the task. Theoretical knowledge is apparent and applied to concrete situations as the student successfully draws conclusions based on the investigations.

Score Point 3: Proficient

The student demonstrates a broad understanding of coordinate geometry. The student uses precise and appropriate language most of the time. Theoretical knowledge is apparent and applied to concrete situations as the student attempts to draw conclusions based on the investigations.

Score Point 2: Apprentice

The student demonstrates an understanding of coordinate geometry. The student uses mathematical reasoning most of the time as well as appropriate mathematical language some of the time and attempts to apply theoretical knowledge to the task, but may not be able to apply conclusions based on the investigations.

Score Point 1: Novice

The student demonstrates some understanding of coordinate geometry but is unable to complete the task. The student uses little mathematical reasoning or appropriate mathematical language. Theoretical knowledge may appear weak and many responses may be illogical as directions are followed incorrectly.

Score Point 0: Unsatisfactory

The student fails to make an attempt to complete the task and the responses only restate the problem.

Chapter 13

Quick Warm-Up 13.1

1. 50%   2. 62.5%   3. $\frac{3}{20} = 15\%$

Lesson Quiz 13.1

1. $\frac{18}{35}$ 2. $\frac{17}{35}$ 3. $\frac{1}{5}$ 4. $\frac{6}{11}$ 5. $\frac{2}{11}$ 6. $\frac{5}{11}$

7. $\frac{5}{36}$ 8. $\frac{1}{3}$

Quick Warm-Up 13.2

1. $\frac{1}{6}$ 2. $\frac{1}{2}$ 3. $\frac{1}{6}$ 4. $\frac{1}{2}$ 5 $\frac{1}{36}$ 6. $\frac{1}{9}$

Lesson Quiz 13.2

1. 3, 6, 9, 15, 18   2. 6, 12, 18

3. 6, 12, 18   4. 3, 6, 9, 12, 15, 18

5.

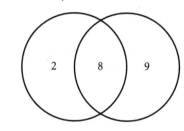

Blue eyes   Blond hair

2   8   9

14

6. 19   7. 14   8. $\frac{1}{2}$

Quick Warm-Up 13.3

1. 2   2. 4   3. 8   4. 6   5. 36

Lesson Quiz 13.3

1. 21

2. 6

3. 360 ways

Answers

Mid-Chapter Assessment—Chapter 13

1. d   **2.** a   **3.** c   **4.** d   **5.** $\frac{1}{8}$ **6.** $\frac{1}{2}$ **7.** 36

8.

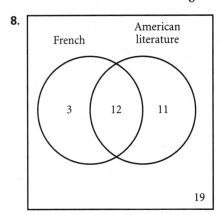

Quick Warm-Up 13.4

1. $\frac{1}{13}$ **2.** $\frac{1}{4}$ **3.** $\frac{4}{13}$ **4.** $\frac{1}{52}$ **5** $\frac{1}{6}$ **6.** $\frac{1}{2}$

7. $\frac{2}{3}$ **8.** $\frac{1}{2}$

Lesson Quiz 13.4

1. $\frac{25}{144}$ **2.** $\frac{1}{9}$ **3.** $\frac{5}{48}$ **4.** $\frac{1}{36}$ **5.** $\frac{1}{6}$ **6.** $\frac{1}{3}$

7. $\frac{1}{6}$

Quick Warm-Up 13.5

1. $\frac{1}{6}$ **2.** $\frac{3}{20}$ **3.** $\frac{31}{60}$ **4.** $\frac{29}{60}$ **5.** $\frac{19}{60}$ **6.** $\frac{31}{60}$

Lesson Quiz 13.5

1. 11   **2.** 5   **3.** $\frac{3}{10}$

Chapter Assessment—Form A— Chapter 13

1. a   **2.** b   **3.** b   **4.** c   **5.** a   **6.** d   **7.** b
8. a   **9.** c   **10.** d   **11.** c   **12.** a   **13.** c

Form B

1. $\frac{1}{5}$ **2.** 0   **3.** $\frac{2}{5}$ **4.** $\frac{7}{10}$

5. Answers will vary. The description of how to simulate one trial and the number of trials that will be simulated must be stated.

6. $\frac{1}{36}$ **7.** $\frac{2}{3}$ **8.** 0   **9.** $\frac{4}{9}$ **10.** $\frac{33}{36}$

11. 1, 3, 5, 7, 9   **12.** 4, 8   **13.** none

14. 1, 3, 4, 5, 7, 8, 9

15.

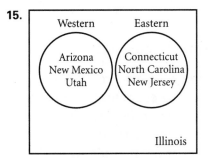

16.

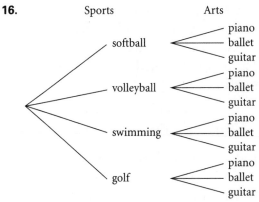

17. 168   **18.** $\frac{1}{2}$ **19.** $\frac{1}{2}$ **20.** $\frac{2}{5}$ **21.** $\frac{1}{5}$

22. $\frac{1}{36}$ **23.** $\frac{1}{4}$ **24.** $\frac{1}{9}$ **25.** $\frac{1}{9}$ **26.** $\frac{4}{9}$

Alternative Assessment—Form A— Chapter 13

1. 60% rock; 40% rap.

2. 4 out of 10 times

3. 2— rock or rap

4. Possible answer: A random-number generator can be used to simulate the problem.

Answers

5. Possible answer: Use a computer, a calculator, or spinner.

6. Answers will vary. The description of how to simulate one trial and the number of trials that will be simulated must be given.

7. The results of a large number of trials are recorded.

Score Point 4: Distinguished

The student demonstrates a comprehensive understanding of experimental probability. The student uses perceptive, creative, and complex mathematical reasoning as well as precise and appropriate mathematical language throughout the task. Theoretical knowledge is apparent and applied to concrete situations as the student successfully draws conclusions based on the investigations.

Score Point 3: Proficient

The student demonstrates a broad understanding of experimental probability. The student uses precise and appropriate language most of the time. Theoretical knowledge is apparent and applied to concrete situations as the student attempts to draw conclusions based on the investigations.

Score Point 2: Apprentice

The student demonstrates an understanding of experimental probability. The student uses mathematical reasoning most of the time as well as appropriate mathematical language some of the time and attempts to apply theoretical knowledge to the task, but may not be able to apply conclusions based on the investigations.

Score Point 1: Novice

The student demonstrates some understanding of experimental probability but is unable to complete the task. The student uses little mathematical reasoning or appropriate mathematical language. Theoretical knowledge may appear weak and many responses may be illogical as directions are followed incorrectly.

Score Point 0: Unsatisfactory

The student fails to make an attempt to complete the task and the responses only restate the problem.

Form B

1. 6 men; 11 women   2. 17   3. 7   4. 8
5. 11   6. a. $\frac{9}{16}$ b. $\frac{15}{32}$ c. $\frac{7}{32}$ d. $\frac{21}{32}$

Score Point 4: Distinguished

The student demonstrates a comprehensive understanding of theoretical probability. The student uses perceptive, creative, and complex mathematical reasoning as well as precise and appropriate mathematical language throughout the task. Theoretical knowledge is apparent and applied to concrete situations as the student successfully draws conclusions based on the investigations.

Score Point 3: Proficient

The student demonstrates a broad understanding of theoretical probability. The student uses precise and appropriate language most of the time. Theoretical knowledge is apparent and applied to concrete situations as the student attempts to draw conclusions based on the investigations.

Answers

Score Point 2: Apprentice

The student demonstrates an understanding of theoretical probability. The student uses mathematical reasoning most of the time as well as appropriate mathematical language some of the time and attempts to apply theoretical knowledge to the task, but may not be able to apply conclusions based on the investigations.

Score Point 1: Novice

The student demonstrates some understanding of theoretical probability but is unable to complete the task. The student uses little mathematical reasoning or appropriate mathematical language. Theoretical knowledge may appear weak and many responses may be illogical as directions are followed incorrectly.

Score Point 0: Unsatisfactory

The student fails to make an attempt to complete the task and the responses only restate the problem.

Chapter 14

Quick Warm-Up 14.1

1. -11 2. -5 3. 1   4. 64   5. 24

Lesson Quiz 14.1

1. Function; each temperature is paired with exactly one wind speed, and no two pairs have the same first coordinate.

2. Not a function; -1, 0, and 1 all have the same first coordinate.

3. Function; any vertical line will intersect the graph at most once.

4. domain: $\{-3, -2, -1, 0, 1, 2, 3\}$
   range: $\{0, 1, 4, 9\}$

5. -1

Quick Warm-Up 14.2

1. 12; 17   2. 9; 14   3. 9; 4   4. 3; 8   5. 3; 2

Lesson Quiz 14.2

1. The vertex is translated 3 units to the left of the origin and 2 units down; $(-3, -2)$.

2. Vertical translation; Every y-value of the function is shifted 1 unit up.

3. parent function: $y = |x|$; stretched by a scale factor of 2 and reflected through the x-axis.

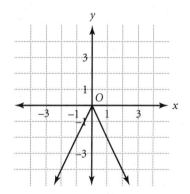

4. parent function: $y = x^2$; translated 2 units to the right of the origin and reflected through the x-axis.

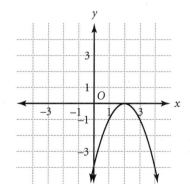

Answers

Quick Warm-Up 14.3

1. 16   2. 4   3. 48   4. 36   5. 12   6. 768

7. 48   8. 0

Lesson Quiz 14.3

1. $y = x^2$

2.
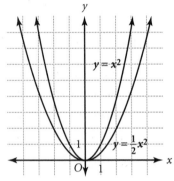

3. The y-values of the graph of $y = \frac{1}{2}x^2$ are one-half of the y-values of the graph of $y = x^2$.

4.
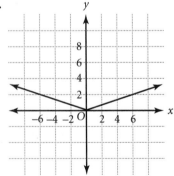

5.
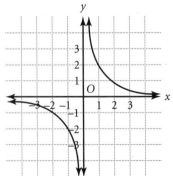

6. The y-values of $y = \frac{1}{5}x^2$ are one-fifth of the y-values of the parent function.

Mid-Chapter Assessment—Chapter 14

1. d   2. c   3. c   4. c

5. domain: {0, 1, 2}
   range: {5}

6. domain: all real numbers
   range: all real numbers greater than or equal to -1

7. 19   8. 1   9. 9   10. $1\frac{1}{2}$ 11. $y = x^2$

12.
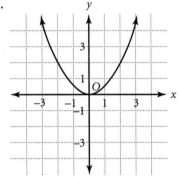

13. stretched vertically by $\frac{1}{2}$

Quick Warm-Up 14.4

1. 30; -30 2. 36; -36 3. 6; -6

4. 64; -64 5. 0.5; -0.5

Lesson Quiz 14.4

1.

x	-2	-1	0	1	2
$f(x)$	$\frac{1}{2}$	1	undefined	-1	$-\frac{1}{2}$

2.
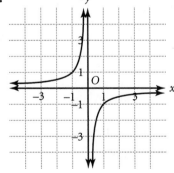

Answers

3. The graph is turned upside down.

4. $y = -x^2$

Quick Warm-Up 14.5

1. vertical translation 3 units down

2. horizontal translation 2 units to the left

3. vertical stretch by a scale factor of 4

4. reflection through the x-axis

Lesson Quiz 14.5

1. Sample answer: Stretch the parent function $y = x^2$ vertically by a scale factor of 2, then reflect the graph through the x-axis, then translate the graph 1 unit to the right.

2.

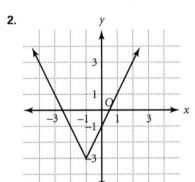

3.

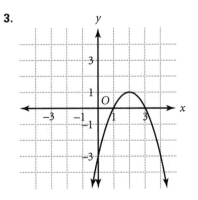

4. $(-2, -6)$

5. $(0, 5)$

6. $(-4, 12)$

7. $(-2, -2)$

Chapter Assessment—Form A— Chapter 14

1. b   2. d   3. a   4. c   5. c   6. a   7. a

8. d   9. b   10. b   11. c   12. d   13. c

14. b   15. b   16. a

Form B

1. A relation pairs any two points. A function is a relation for which no two points have the same coordinates.

2. $(-2, 3)$ 3. not a function   4. function

5. $-\frac{3}{2}$ 6. $\frac{1}{2}$ 7. -2 8. 1   9. 6   10. $-1\frac{1}{2}$

11. The graph is turned right side up (a V that opens upward).

12. $y = x^2$

13. $y = |x|$

14. $y = \frac{1}{x}$

15.

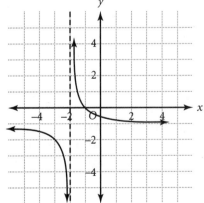

16. a vertical translation 1 unit down followed by a horizontal translation 2 units to the left

17. stretch; 2   18. compression; $\frac{1}{2}$

19. stretch; $\sqrt{3}$ 20. neither; 1

Answers

21.

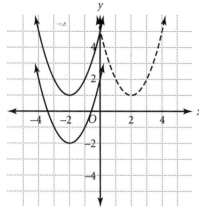

22. $(-3, -1)$

23. compressed   **24.** translated

25. translated   **26.** reflected

Alternative Assessment—Form A— Chapter 14

1.

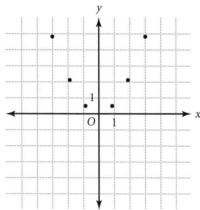

2. Experiment with various transformations of a parent function until you match the graph of the parent function.

3. The parent function is $y = x^2$.

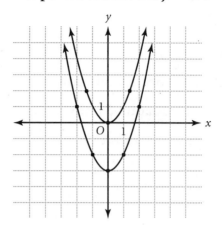

4. Answers may vary. Sample answer: A point on the parent function can be compared with the corresponding point on the function in order to observe that each point has been translated down 3 units.

5. The transformation applied to $y = x^2$ is a vertical translation of -3.

6. $y = x^2 - 3$

Score Point 4: Distinguished

The student demonstrates a comprehensive understanding of basic transformations of functions. The student uses perceptive, creative, and complex mathematical reasoning as well as precise and appropriate mathematical language throughout the task. Theoretical knowledge is apparent and applied to concrete situations as the student successfully draws conclusions based on the investigations.

Score Point 3: Proficient

The student demonstrates a broad understanding of basic transformations of functions. The student uses precise and appropriate language most of the time. Theoretical knowledge is apparent and applied to concrete situations as the student

Answers

attempts to draw conclusions based on the investigations.

Score Point 2: Apprentice

The student demonstrates an understanding of basic transformations of functions. The student uses mathematical reasoning most of the time as well as appropriate mathematical language some of the time and attempts to apply theoretical knowledge to the task, but may not be able to apply conclusions based on the investigations.

Score Point 1: Novice

The student demonstrates some understanding of basic transformations of functions but is unable to complete the task. The student uses little mathematical reasoning or appropriate mathematical language. Theoretical knowledge may appear weak and many responses may be illogical as directions are followed incorrectly.

Score Point 0: Unsatisfactory

The student fails to make an attempt to complete the task and the responses only restate the problem.

Form B

1. 1. Determine the parent function.
   2. Determine whether there are stretches and/or reflections of the graph of the parent function.
   3. Determine whether there are translations of the graph of the parent function.

2. The graph is a reflection of the graph $y = |x|$ through the x-axis, shifted 3 units to the left and 1 unit up.

3.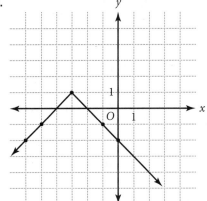

4. The order of transformations for this graph makes no difference.

5. In general, transformations are not commutative.

Score Point 4: Distinguished

The student demonstrates a comprehensive understanding of how the combinations of transformations are performed. The student uses perceptive, creative, and complex mathematical reasoning as well as precise and appropriate mathematical language throughout the task. Theoretical knowledge is apparent and applied to concrete situations as the student successfully draws conclusions based on the investigations.

Score Point 3: Proficient

The student demonstrates a broad understanding of how the combinations of transformations are performed. The student uses precise and appropriate language most of the time. Theoretical knowledge is apparent and applied to concrete situations as the student attempts to draw conclusions based on the investigations.

Answers

Score Point 2: Apprentice

The student demonstrates an understanding of how the combinations of transformations are performed. The student uses mathematical reasoning most of the time as well as appropriate mathematical language some of the time and attempts to apply theoretical knowledge to the task, but may not be able to apply conclusions based on the investigations.

Score Point 1: Novice

The student demonstrates some understanding of how the combinations of transformations are performed but is unable to complete the task. The student uses little mathematical reasoning or appropriate mathematical language. Theoretical knowledge may appear weak and many responses may be illogical as directions are followed incorrectly.

Score Point 0: Unsatisfactory

The student fails to make an attempt to complete the task and the responses only restate the problem.

DATE DUE

MAR 27 2003

JUN 11 2004

FEB 2 0 2007

GAYLORD

PRIN-